Ins & Outs of Ferry Flying

Dedication

In memory of Ruth Crawford Downie, 1915-1977, who flew many ferry flights and waited out many others. May she have a good flight.

Other TAB Books by Don Downie:

2208 *Cockpit Navigation Guide*

Ins & Outs of Ferry Flying

by Don & Julia Lee Downie

MODERN AVIATION SERIES

TAB BOOKS Inc.

BLUE RIDGE SUMMIT, PA. 17214

FIRST EDITION

FIRST PRINTING— JUNE 1980

Printed in the United States of America

Library of Congress Cataloging in Publication Data

Downie, Don.
Ins and outs of ferry flying.

Includes index.
1. Airplanes—Ferrying. I. Downie, Julia, joint author. II. Title.
TL711.F47D68 629.132'52'068 80-16310
ISBN 0-8306-9936-8
ISBN 0-8306-2280-2 (pbk.)

Cover Photo of Piper's new Seminole light twin, courtesy Piper Aircraft Corp.

Contents

Acknowledgements

No book is assembled under one cover without the cooperation of many people. We would like to thank the editors of *AOPA Pilot* magazine whose copyright material is excerpted with the permission of Ziff-Davis Publishing Co., which presently has all rights to the *Skyways* title; the editors of *Plane & Pilot; Air BP* in London; *Soaring* magazine published by the Soaring Society of America; and the Pasadena, CA, *Star News* for their permission to reprint personal experiences that led to *Ferry Pilot.*

Special appreciation goes to old-time aviation editors who cooperated to make many of these continuing ferry flights possible; particularly to Doris Ahnstrom who was editor of *Skyways,* the late Albert Lewis with Condé Nast publications in New York City, and Charles P. Miller, for many years managing editor of *AOPA Pilot* magazine. Then add the names of editors John W. R. Taylor, Jack Leggatt and Bob Said. One of the best proven ways to arrange a ferry flight is to assure an unbiased report of the trip in print; these editors and many others helped.

David Hatfield, historian, furnished background information. Assistance from the many manufacturers came in the form of check rides, accuracy review of editorial material, and service above and beyond the norm.

We should note that the experiences, events and techniques shared with the reader are not meant as instructions or directions on how to fly. In no way are they intended to indicate the best or only way to handle situations in flight or on the ground. You're strictly on your own.

Delivering new airplanes, shuttling used ones and running local errands by air is all ferry flying of one sort or another. The ferry pilot has the singular advantage of making his own decisions, plotting his route, picking weather that fits for him and his equipment, and stopping at places of interest as time permits. There isn't the nervous strain, responsibility and repetition of teaching circuits and bumps. However, you can't afford to stick your neck out and bend an airplane. Without exception, applications for ferry flying list: "Have you ever had an application for hull and liability insurance declined?" and "Have you had any aircraft accident while acting as pilot?" or "Have you ever been cited for violation of civil or military aviation regulations? If so, provide details."

The following pages cover many air miles and many hours over some of the most spectacular country in the world. With a couple of notable exceptions, the trips have been a delightful change-of-pace from the 40 keys of a hot typewriter, the shadowed confines of a photographic darkroom or the apprehensive monotony of a big city freeway.

It will be interesting to see where that next ferry flight goes. We hope that yours might be as enjoyable as ours.

Don and Julia Lee Downie

Chapter 1

So You Want To Be A Ferry Pilot

There was a snowstorm outside the barbecue diner in Liberal, Kansas. It was the fourth of May and the weather was unseasonably socked-in. Inside, four ferry pilots and their crews were "hangar flying" and waiting for better weather.

Tied securely on the airport ramp were our Grumman Tiger, a Piper Tomahawk, a Cessna 182RG and an Engstrom Helicopter. Each crew had a different reason for ferrying a new airplane. Julia and I were doing a magazine report on the new Grumman. Wynn Kinner was taking the Tomahawk west so that he could visit friends throughout the country, save transportation costs and deliver the trainer for a friend. The new 182RG had just been purchased by the pilot who was returning with it to Phoenix, AZ. The helicopter pilot and his wife were delivering the 'chopper' to the San Francisco area. The pilot was taking his two-week vacation from the Detroit City Police force and showing his wife the United States as an indoctrination trip.

What Is Ferry Flying?

Ferry flying is one of the most satisfying of aeronautical experiences. There is usually no student instruction or the monotony of traffic-pattern training. The pilot is free to choose his route and timing. The equipment is frequently brand new and retains that new-airplane smell that will never be duplicated, despite "new" sprays used on some older aircraft. Whatever the motive for the delivery, each ferry pilot has the sole responsibility for a safe delivery both of the airplane and his own hide(Fig.1-1)

Early the next morning, skies were clearing in Liberal and the four aircraft departed for their varied destinations. We talked on the radio for a distance. We compared notes with Kinner as we both topped off in Albuquerque. Then we were off again to do our individual "things."

There are several motivations to become a ferry pilot, either part-time or as a professional. Probably the number one reason is pure love of flying, with expenses paid, to see the United States. Some new pilots use it as a time-builder for either executive or airline flying. Instructors will frequently make a delivery for their own school and take a student along to gain cross-country experience. Some, like the author, combine a love of flying with research for magazine articles on new aircraft, places to fly and a review of the Air Traffic System. Frequently, owners save the ferry cost and pick up their own new aircraft.

Larger, complex aircraft—twins, jets and special-purpose airplanes—are usually picked up at the factory by company personnel of the client. If the new ship differs from the existing model, frequently the Chief Pilot and Chief Mechanic of the purchasing company will jet to the factory and take a concentrated ground school in the various systems of the new airplane.

Special ratings are required for twins, seaplanes and helicopters. If you're on the end of a tow line, a glider rating is required. A sampling of what various organizations want in pilot experience will be found in subsequent chapters (Fig. 1-3).

Most organizations picking up complex aircraft where there is a high-dollar investment will require a commercial license and sometimes an instrument rating. The latest change in FAR's (Federal Air Regulations) now lumps the commercial and instrument training at accredited schools.

FAA rules and regulations change. At the moment, a student pilot can legally make a one-day ferry flight if he has been briefed and signed off by his instructor. In fact, a qualified instructor and one of his students can really take two airplanes all the way across the country if the CFI gives his student a briefing every day and signs him off for that day's flight. However, a private license is considered minimum for VFR day cross-country deliveries.

If your destination happens to be San Francisco International, as it was for us recently with a four-placer, you need to be knowl-

Fig. 1-1. There's nothing like the brand new smell and touch of a factory-fresh airplane. This Cessna Skyhawk comes back from a test-hop at the factory delivery center in Wichita, KS, prior to heading toward a far-away destination.

Fig. 1-2. The winding Arkansas River passes near downtown Wichita as a new Cessna 175, flown by the author, heads for California. Careful fuel and oil checks should be made on the initial leg of a delivery, particularly if there are mountains and deserts before you reach your final destination.

Fig. 1-3. Twin-engine deliveries cut flight time but require adequate experience and currency in the particular type of plane you pick up. new Beech Baron heads for Albuquerque, NM, after leaving Wichita with factory pilot Bob Buettgenbach at the controls. Photo was taken from a Beech Sierra headed for Van Nuys, CA.

edgeable about all air traffic control procedures and, in particular, the predictability of wake turbulence and what to do about it. This is true of all big airports and TCA's. If it is a clear afternoon at SFO and you're cleared to land on 28L "behind the National DC-10 at your nine-o'clock position," you'd better stay on the upwind side of the ILS and let that rugged wake turbulence drift away. Then contact the tower and request "a long landing." The tower is accustomed to such requests and merely gives a non-committal, "Roger, cleared to land." If you try to come in "trailing" the heavy DC-10 down final approach, you can expect a non-unit course in low-level survival aerobatics. It could ruin your whole day. Either know how to mix in with the "heavies" or stay out of the air-carrier hub airports (Fig. 1-4).

Ferrying Expenses

Today's commercial license usually incorporates an instrument rating. As flights get longer, with IFR and night operations included, the fly-for-hire rating is most desirable. While reimbursed expenses are not considered as "commercial" flying, many insurance companies want that rating before covering a new airplane on delivery.

If you ferry a ship for expenses only, it is prudent for income tax purposes to have your expense account and your reimbursement check say "for expenses only" so you don't have to pay a tax on the non-income you receive.

More often than not, new pilots actually do some ferrying without really knowing it. If there's a ship to pick up across town, that delivery is really a short ferry flight. Julia Downie had only 18 hours solo time when she and her instructor, Bob Cain, went out dual from San Jose, CA, some 50 miles to Salinas to pick up Cain's Cessna 150 from a paint shop. When Julia, complete with student permit at that time, came back on what was really a cross-country solo, she was also involved in a ferry flight.

Local FBO's, maintenance shops and used-aircraft dealers frequently draw on the abilities of their own local pilots to pick up or deliver aircraft. In most cases, this is done on a non-revenue basis. The pilot either ferries across town in an available ship, takes a commuter airline, or whatever. Usually, the pay-off is in flight experience—time building, if you will.

In some cases, new pilots are eager enough to add to their log books that they will pay their own expenses just to get the

Fig. 1-4. Many ferry flights include landings at large air carrier airports. Here in San Diego, a new Piper Cherokee waits to take off as a PSA Boeing 727 approaches for a landing. Be sure that you are familiar with the latest TRSA or TCA requirements for these larger airports and that your plane has the required transponder/encoder equipment aboard.

flying experience. Each ferry flight has its own particular arrangement. When you get into the longer deliveries, the flight can be arranged for a flat fee; if the pilot wants to carry canned beans and a sleeping bag, that's up to him (Fig. 1-5).

Commercial ferry services, however, pay either by the flight hour or by the trip. Since this is a fly-for-hire operation, commercial and, usually, instrument ratings are required. Almost all such services have pilots who are individual contractors, although the ferry service usually handles liability insurance.

Your Travel Bag

What should you take along on a ferry flight? The answer depends a great deal on the route and time of the year. There is no such thing as a standard survival kit unless you want to haul around a 50-pound sealed container.

You'll want your own personal toiletry kit that includes whatever legal medication you might need. A doctor's prescription for broad-spectrum antibiotics is good insurance in case you come down with a cold, or whatever, while on the trip. Lomotil, a prescription drug in the United States, but over-the-counter in

Latin American countries, is an excellent quick-fix for diarrhea. Suntan lotion or cream will save your face: dark glasses help during that afternoon flight westbound, and noise-canceling headsets are being more and more recommended as hearing protection. Personally, I carry a Sigtronics headset and boom mike (with a plug-in Sigtronics intercom package if someone is going to fly with me). It cuts down on cabin noise and fatigue and also makes it much easier to talk in flight, particularly in tandem trainers.

If your trip is in the summer and covers any extensive desert area, stop at the supermarket just before you pick up the plane and get a gallon or two of water—you don't really want to pack it along on the airline—and whatever "munchies" you want in flight. Some pilots use chocolate bars, raisins, crackers or a package of cold cuts. Chewing gum helps others.

For a winter trip, a reflective "space blanket," weighing only a pound or so, provides excellent warmth in case of an off-airport landing.

Fig. 1-5. How do you get there from here? Similar signs at the Tradewind Airport in Amarillo, TX, and Salt Lake City, UT, (right) show airline distances to some places you might never have heard about.

Fig. 1-6. Ferry pilots frequently extend flight time into darkness hours. Most pilots with new single-engine aircraft prefer to have a few hours of break-in time before launching out at night. Here a Cessna approaches the Los Angeles, CA, area at night.

Cold-weather clothing should include gloves, cap and heavy wool socks. Flannel pajamas can double for "long johns" on the initial cold part of the trip from places like Alexandria, MN, or Osceola, WI.

Nylon tie-down ropes are always useful and take up little space. A hunting knife, and perhaps a short-handled ax, might be very helpful. Personally, I take a small ax and six aluminum extrusions, three or four feet in length, for tiedowns. Three canvas sacks will do the same job, if you fill them with rocks or sand.

A waterproof pack of matches is a must and a simple signaling mirror should be part of any pilot's map case, even if you are only planning to fly across town.

Take at least two flashlights. Even if you plan only day VFR, the flashlights provide a protection against the unexpected delay, head winds, a break in the weather where you want to fly until the late dusk, or whatever (Fig. 1-6).

Of course, you're going to need charts—even for a ferry flight from a shop across town. The larger the stack of charts, the greater the detail. Some pilots can fly well with only the four jet planning charts that cover the entire United States. Others get along nicely with the WAC scale. Personally, I prefer Sectionals when they're available, even though the bulk, weight and costs are greater.

Even if you plan only a VFR flight, low-altitude enroute charts are useful. IFR in a brand-new airplane, or one that is brand new to you, is a question of personal preference. The new ships all

Fig. 1-7. Betty Jo Reed carefully puts all her VFR Sectionals together prior to taking off in a Mooney 21 during a "Powder Puff" competition sponsored by the '99's. It takes all kinds of charts to handle a ferry flight. Sectional charts have greater detail for a VFR flight, although some ferry pilots have been forced to use road maps and even the Best Western Motel chain surface booklet that covers the entire country.

have a complete flight test before you get into them, but on a simple airplane like a Cessna 150, total time will usually be less than one hour. Predictably, there is more chance of a malfunction of airframe, engine or avionics in the first few hours than later in the life of the airplane. Some dealers request VFR only on a delivery and write that into their insurance package. Others with more sophisticated equipment leave it up to the pilot. Personally, I never go IFR on the first leg out of the factory, and I've waited as long as five days in the Ramada Inn in Wichita for a break in the weather; but with a Cessna 150, it was well worth it.

If all else fails, get a package of state road maps. They don't show nav/aids or some altitudes, but names of the towns are there along with main highways. Some pilots, this one included, have flown short legs of a delivery flight with the handy-dandy Best Western Motel booklet of state maps. Most smaller towns carry their name on the water tower and perhaps printed on the airport runway. In wide-open country, there's nothing but the interpretation of the FAR's that precludes dropping down low enough to read

Fig. 1-8. Portable transceivers and temporary nav-com installations operate with varying degrees of success. Rick Skogland, President of California Air Radio at Cable, CA, checks over an Escort 110 that he has adapted for temporary installation on a ferry flight.

Fig. 1-9. A classic restored Monocoupe prepares for a cross-country delivery to its new owner. The Monocoupe was restored by Jim Dewey of Santa Paula, CA, who was flying this ship at sunset over Ventura, CA. The Channel Islands are visible in the background. If you plan to fly a taildragger like this, be sure that you are familiar with the characteristics of takeoff and landing that are unique to this configuration.

the mileage ahead to the next big town. This poor pilot's DME can be a great help (Fig. 1-7).

Several commercial services use portable tranceivers (Fig. 1-8), and the larger FBO's which bring out any number of NORDO aircraft will usually have some sort of a portable either in their shops or a "something" they can rent from a nearby radio shop. My personal experiences with portable transceivers have ranged from poor to useless. In the first place, you have to carry them aboard the airline or carry a bulky, crush-proof packing crate. Then when you get your new airplane, you have to "Mickey Mouse" some sort of an antenna arrangement and either patch into the aircraft electrical system or hope that the batteries on the portable are new. On battery-powered com or nav/com units, bring a spare power unit.

Many of these units are bulky and should be secured carefully by seatbelt or nylon rope. Otherwise, they might try to join you in turbulent air.

Regency once made a small 12-channel transceiver that worked on occasion when the spike antenna was merely held near, or out, an open window. However, it had no navigation capability. The backup of a communications set makes your ferry flight easier since you have access to almost all airports. Add a portable VOR receiver and things get even less complex. However, if you start out in a NORDO unit and know right from the beginning that you're going to work pilotage to smaller airports, spend a little money on weather and flight planning phone calls and keep a friendly relationship with interstate freeways; it really isn't all that bad.

There's one relatively new type of quasi-navigation you shouldn't forget. When the weather is right, jet airliners lace a pattern of contrails across the skies—usually in a direct line between large towns. If you are familiar with the area, you can use these as navigational aids. It is easy to locate the Grand Canyon, for example, from a hundred miles away since that's the only place I know of where these contrails consistently make "S" turns.

So, welcome to the sometimes wild, usually wonderful, world of ferry flying (Fig. 1-9). We'll take a few trips—some old and some brand new. We'll fold carefully into some tiny puddle-jumpers and stretch out in larger, single-engine aircraft and light twins. We'll prop-jet and pure-jet. We'll recount other pilot's "ferry-interesting" experience and repeat some of our own. So, let's go flying!

Chapter 2

How To Get Started

One of the most satisfactory fly-for-fun, transportation and magazine-article-research arrangements I've had over the past years has been with Santa Paula super-salesman, Mike Dewey. To date, I've ferried 17 new aircraft for him. All but one were the "tail-dragging" Citabrias and many were without radios. My on-going financial arrangement with Mike has been strictly expenses only. He picks up the air coach fare to the factory, reasonable overnight motel and food charges and gas and oil on the new airplane. My profit is in getting to some place I want to visit without any out-of-pocket expenses, enjoying the wild blue yonder or picking up material and photos for a magazine assignment.

What a Dealer Looks For

Mike asks three things from his part-time ferry pilots:

Time. You must have no time schedule commitment when you have to be back home. "Any pilot who doesn't have plenty of time to make the trip back from Osceola, Wisconsin—and it never takes less than 20 flying hours—just doesn't go and pick up a ship for me. If he says, 'Fine, I'll take the trip, but I've gotta be back next Monday,' it's no dice. Between weather delays, occasional maintenance or lack of fuel in the short-ranged 26-gallon trainer tanks, I don't want a pilot who is in a desperate hurry."

Proficiency. You must be fully proficient in a taildragger (Fig. 2-1). The taildragger has a built-in longitudinal control problem

Fig. 2-1. Taildragger pilots take a certain pride in their ability to keep ahead of what are now "unconventional" aircraft. A California club is limited to those who frequently fly taildraggers. Aircraft sales organizations which handle tailwheel aircraft report an increasing problem in finding pilots who can fly this type of equipment.

(ground loop possiblity). Add icy runways along the route from Wisconsin in the winter and you have a rather challenging situation.

Navigational Ability. You must be able to navigate without radios and nothing more than a Sectional chart and a road map. "Whenever we send a new pilot back to the factory, we'll take him out as part of our flight check and make sure he can read maps or charts. We want a 'highway guy' rather than one who goes from

VOR to VOR. Now, almost all our ships come out with some sort of a radio, but over the past years, only 40 percent of that total has been radio-equipped. In addition, the reliability of cheap radios has never been the best. We find that the average 100-to 500-hour pilot who learned to fly at a typical big-city flight school can't fly from A to B unless all the avionics work. They're really kinda half lost and don't even know it most of the time. It's only their familiarity with their own local area that gets them back home."

"You know, it's getting harder and harder to find pilots with those qualifications," explained Dewey as we kicked around this manuscript. Dewey's "Screaming Eagle Aviation" sells an average of 45 Citabrias per year and is one of the three top sales organizations in the country for this model. Usually, 40 percent of the new aircraft are picked up by the owner and Dewey arranges for the remainder.

"We always send back someone we know and preferably someone we have trained here," explained Mike. "We like a maintenance background if we can get it, but that isn't really essential. These new production airplanes are so reliable that we seldom have a mechanical delay any more, with the possible exception of avionics (Fig. 2-2).

"Our insurance requires a minimum of 500 hours with at least 25 in type. The only pilot we have used consistently who has less than that total is John Peterson who works for us on the flight line. He's also my nephew. I taught him to fly and he's using the ferry trips to build up his background of experience."

In preparing new pilot Peterson for ferry flying, he was sent out on one delivery with veteran Jim Dewey, Mike's father, in the same airplane. This enroute instruction was invaluable to the new pilot. Then three trips followed in a "flight of two" where a seasoned ferry pilot called the shots on weather, routing, fuel stops and overall planning. Only after this briefing did Peterson go it alone.

Dewey explained that all new aircraft carry a $500 deductible insurance policy, and a commercial pilot is not specifically required. "However, a minor 'prang job'—even just a little bit—can easily cost the FBO $4000. You have to bring the pilot home on the airline after he gets in trouble, have the ship repaired in someone else's shop at their labor rate—not yours—send back another pilot to complete the trip, and then sell the repaired airplane as a used ship. So that $500 limit is just the beginning."

Over the years, Dewey has had only two new aircraft wrecked. One was in a hangar in Texas that was hit by a tornado.

Fig. 2-2. It's hard to do, but some pilots find it possible to fold the nose gear of a Cessna 150. This photo, taken at Brackett Field in Southern California, was the result of "wheelbarrowing" in trying to force the nose wheel on the ground at too high a landing speed. Ferry pilots should be completely familiar with their new aircraft or get a thorough check ride.

The other was dunked in the Yellowstone River following some low-flying by the pilot.

"Normally, we wouldn't use him again, but he had brought out 30 aircraft before this accident," said Mike. "The Citabria was covered by insurance except for the deductible, which ran to about $2,000. The pilot involved insisted that he pay all costs, which he did, and we sent him out after another airplane. He hasn't been in trouble since."

"Some of our ferry flights are made by airline-types who make a vacation out of the trip back. They have the advantage of a space-available airline pass that either cuts my ferry cost or gives the pilot a little profit. We had the son of one Continental Airline captain, John F. Solomon, who made 30 deliveries for us before he was 21 years of age. Only after that flight experience could John qualify for the air carriers. He is now flying out of Honolulu for Continental."

Aircraft dealers have a high exposure to accidents because they buy not only new, but also used, aircraft and must shuttle these ships from one airport to another. The chances of a mechanical failure are greater during the first few hours of flight. Some of the used aircraft are being sold because of anticipated maintenance problems. For this reason, Mike Dewey has built up a portable shoulder harness that goes right in his map case.

"I seldom fly without a shoulder harness," he explained (Fig. 2-3). "I've done some consulting for attorneys on the "crashworthiness" of aircraft with and without restraint systems other than the lap belt. If I were to go out and pick up an old Cessna 195 from a dirt strip near Boise, for example, I'd be in a pack of trouble if the owner neglected to tell me that the carburetor was about to fall off—thus, the shoulder harness."

Herman L. Martin, veteran Piper Dealer of Walla Walla, WA—actually, his neat-as-a-pin, 2600-foot airport is in nearby College Place—has some very definite ideas on ferrying. And he should know. He has sold over $500,000 in new aircraft any year, and has been Piper's No. 2 salesman nationwide. The walls of his office are covered with other sales effort plaques.

"We don't use ferry services any more," he stated flatly. "Some ferry services send their pilots out on a flat-fee basis. One of the ways the pilots will save a couple of bucks is to run the engines just as lean as possible.

"Recently, we traded a new Cherokee 180 with another dealer because of a paint job. The other ship had been brought out

Fig. 2-3. Super salesman Mike Dewey shows his personal ferry-flight kit to pilot-pharmacist Jack Switzer in preparation for a Bellanca Decathlon delivery from the factory. Dewey carries a small kit of tools, add-on shoulder harness, two flashlights, charts and a current copy of AOPA's Airport Directory to help pick suitable landing spots with restaurants, nearby motels, etc.

by a ferry service, and it was just 100 hours before we had to 'top' the engine because of burned valves (Fig. 2-4).

"We'll send back someone we've trained and pay all their expenses. Some of our 140's and other trainers have been brought out from Vero Beach by private pilots with only 100 hours who are working toward their commercial licenses. We instruct these newer pilots never to fly in weather, and we've never had a problem.

"If there's a tail-dragging Pawnee to bring out, or a new twin, one of my two sons will usually make the flight."

Many pilots have never been into a Class I TCA, and some of these will wind up ferrying a ship into a metropolitan hub airport. It's not all that difficult, even if the ship is not transponder-encoder equipped. The FAA has the capability of approving a one-time takeoff or landing for delivery or maintenance where the plane has less than the prescribed avionics. It does take a phone call in advance to get this permission and you have a better chance of receiving approval if you try for non-peak traffic hours (Fig. 2-5).

Fig. 2-4. A Piper Cherokee nears the end of a delivery flight in Southern California (Courtesy Tom Roberts).

On a couple of the no-radio deliveries I have made, it has been desirable to land at tower-controlled airports. You can come in on an FAA VFR flight plan with the notation of NORDO on a traffic-permitting basis, but you should be right on your ETA. Frequently, the busy facility will ask that you land at an airport close in and phone first.

When you advise that your "radio is out," there really can be more than one definition.

It's not only prudent, but it is also a legal requirement to be current in the type of aircraft you're ferrying. Most dealers will require a thorough check-out before delivery, not only for your benefit but to meet insurance requirements. Therefore if you're picking up a particular brand of twin, it is best that you get a check-out even if you have a great many twin hours in your log book.

If your craft is a taildragger, don't try to bluff it out if you've never flown one. Chances are you'll scuff it up on your first or second attempt without some check time.

"Complex aircraft," as defined by the FAA, have all those good things like retractable gear, flaps and controllable props. They present no big problem, but if you've never flown them before, you really do need a check-out—if for nothing more than your peace of mind (Fig. 2-6).

Turboprops and pure jets aren't difficult, but a type rating is usually required to make you legal.

Fig. 2-5. Busy air carrier airports like Burbank will accept no radio or radio-out aircraft on a one-time landing if there is sufficient justification and a telephone call is made in advance for permission. Light gun signals call for waggling the wings. This approval is at the option of the acting tower chief.

In some aircraft you just can't get a check ride. Single-seaters and some homebuilts present this problem. If the single-seater is something like a sleek P-51, you'll want a back-seat check in the black (and blind) hole of Calcutta in an AT-6 with a well-qualified pilot. If the ship is a homebuilt, get a good ground check from the builder with particular emphasis on the fuel system. You can't go far in a powered aircraft if you can't get fuel to the engine.

Cautious taxi testing with taildraggers and experimental homebuilts is a prudent pre-flight. Get the feel of all the controls and just how the ship handles on the ground before you go blazing out into the sky.

Delivery of new airplanes with values from $20,000 for a bare trainer to $100,000 for a well-equipped four-placer—and on to several hundreds of thousand for a complete twin—make ferrying a serious business. However, there is an occasional new plane pick-up that falls into the category of just plain (plane?) fun. It takes that rare combination of an on-time delivery where everything works, a succession of good weather days across the country, friends and relatives available when you can meet with them, shuttle airlines that are on schedule and that flexibility in your own time schedule to permit enjoying this combination to the utmost.

What follows is such a delivery. The report is printed for the first time here, because this trip wasn't made on assignment.

Once Upon a Long Weekend

Little airplanes are more fun than just about anything. When you work with them and report their activities, you must come up with a complex package of numbers: weights, speeds, range, performance and, perhaps, a bad habit or two. Then you analyze the data, put in more numbers and get it into print. The result is informative—but not always pure, unadulterated fun.

Take a long weekend like Labor Day with no note pad—no computer or checksheet—no assignment to report. Just pick up a new airplane on the east coast of Georgia and take a long, leisurely weekend with relatives. Then deliver the new ship on the West Coast and go back to work.

It really doesn't make any difference that the airplane was one of Grumman's sparkling new Tigers. That applies to our just-for-fun weekend only in that the kick-off was in Savannah, GA (Fig. 2-7).

First leg was solo with a destination of Washington, D.C. Flight plan? No way! Circle the field to check the gauges and then

Fig. 2-6. Complex aircraft require pilot qualifications and a check-out. Here's the panel of a Beech Baron photographed in flight somewhere between Albuquerque and the Grand Canyon. Indicated air speed is 183 mph at 8800 feet in turbulent air. The proper en route altitude here would be 8,500 feet.

Fig. 2-7. Tiger Center, Grumman's delivery center at Savannah, GA, was the starting point for this long, leisurely weekend designed strictly for fun.

head up the East Coast. Never been that way before. Cruise the broad beaches, oggle the multi-colored sailboats and barges on the intercoastal waterway.

As time permits, check out the VOR and transponder to get into Washington's TCA. With the "red eye" east and three cups of coffee for breakfast, a lunch stop came early. We'd never before landed at Myrtle Beach, SC. The tower was good, two FBO's vied for our 20 gallons, and one offered a ride to town for something other than a machine-served lunch.

Bummed a ride back from the restaurant with a guy who once made Pipers in Lock Haven. Small world this flying community. Then off, boring small holes more-or-less toward Washington, D.C.

The weather is good and the map shows an intriguing line of sand spits up toward Cape Hatteras. Remember the movie, *Gone With the Wind?* Let's go that way rather than direct on Victor 39. It's less than an hour longer.

Besides, the Sectionals show the "first flight" airport at Kitty Hawk. Never been there either.

Coastal water is clear and spotted with sailboats and yachts. Pick our way around the restricted areas inland of the Cherry Point Marine airport and head for Ocracoke Island and up the narrow sand spit that becomes Cape Hatteras (Fig. 2-8).

Fig. 2-8. Isolated, sandy beaches along the southern Atlantic Coast deserve a close inspection. There is no problem in flying close to the beach as long as you remain at least 500 feet away from people or buildings.

Small auto ferries buzz across the small inlet at Ocracoke Island to help ground-bound vacationers pick their favorite sand dune to park an RV behind. Interesting lattice-work patterns show up on windbreaks that must be designed to divert shifting.

The adequate airport at Hatteras, commemorating a very fine military aviator, Billy Mitchell, is equally empty. We wonder why more people don't fly down for the weekend, but this "far-east" sand spit could be a real stinker when the weather went sour.

The Sectional reminds, "Cape Hatteras National Seashore. Landing aircraft PROHIBITED." All well and good, but the dune buggies were having a ball, and if that new Lycoming had developed the slightest murmur, we'd have worried about the regulations later.

Comes Manteo, Nags Head and then the Wright Monument (Fig. 2-9). Circle once, admire the statue, land, shut down, sign the guest book and walk to the top of the hill where this thing all started. There's one twin Comanche in the parking area, but it

Fig. 2-9. The Kitty Hawk Monument is always a favorite shrine for aviation buffs. The paved First Flight Airport here is 3,000 feet long and within easy walking distance of the monument.

takes off as we climb the hill. One heck of a lot of flying has been logged since then.

We look and admire the grass at Kitty Hawk where the first flights were made. Wish we could have landed there rather than on the paved strip—but it's not really so bad.

Back in the cockpit after a phone call to have Jake pick us up two hours later at Gaithersburg. Gotta stop for fuel. It turned out to be a comfortable "Ma and Pa" operation at Williams, WV, after a helping hand from Newport News Approach over a little area of "1,000 and 3." No great problem.

Sign for the fuel, pay 20 cents to recheck FSS with improving weather reported. Climb up into the murk; latch onto Interstate 95; talk with Richmond Approach, and appreciate the non-standard transponder in this new bird. Most deliveries don't have 'em.

Contact Washington Approach, get talked around downtown and pointed more-or-less toward Gaithersburg. Good thing I've been there before because Approach had us 20 degrees off, but no sweat. It's still a beautiful evening for a long weekend.

Meet our daughter—it's been three or four months. Check the weather to find a good forecast for the next two days, so we regroup and spend the day at the Smithsonian—the new Aerospace Building, naturally. Interesting to see the Wright Brothers' plane after visiting Kitty Hawk the day before.

Touring the Smithsonian is hard on the leg muscles, but well worth the time and effort. Most aviation buffs who have visited the Aerospace Building at the Smithsonian wait in line to see a wide-screen of early days of ballooning with a series of truly spectacular shots of the face of America. Only from the air can this face be appreciated. It was interesting to sit in the packed theater and watch the panorama unfold, knowing that you were really there yesterday, picking your own view of this country from over the nose. Even more intriguing was the knowledge that there were almost 3,000 more miles to go during the remainder of the weekend with no boundaries other than weather, a certain prudence, and a liberal interpretation of a few FAR's (Fig. 2-10).

Found one of my "AOPA Pilot" cover photos on display with a credit to Great Lakes Aviation. Fair enough—it was there.

Next morning, Jake gives us a ride to Gaithersburg and our daughter climbs into the driver's seat. It was her first cross-country trip in a little airplane since she was banged up in the wrong airplane with the wrong pilot at the wrong time. The bones are mended, but we're looking for sweaty palms. None appear and we're happy.

Fig. 2-10. Cloud build-ups like these always make a cross-country delivery interesting. They're never the same on any two trips.

Off into the murk and across those lethal little hills called the Appalachians. Over Cumberland we regroup and circle up through a hole to 8000 on top. There's always some ground contact, so it's a good decision.

I relax in the right seat and watch. Weather falls away as we go west. Surprise! The forecast was correct, but Flight Watch on 122.0 is jammed with pilots calling Washington, Detroit, Pittsburgh and, soon, St. Louis. They really need a discrete frequency for higher-altitude weather information. Maybe we can help do something about it.

We plan to fuel and lunch at one of the "burgs" along the Ohio River. This one is Parkersburg. Weather is perhaps three miles with haze—we call it smog—so we cheat a little and bird-dog in over the VOR. The tower clears us straight in with a report at the river. Still no airport in sight, but soon it emerges from the murk and Dana shoots her first landing in several months. It was interesting.

Phone ahead and head for London, Ohio, to talk with the designer of a tail wheel modification for the nosewheel Cessna 150-172's (Fig. 2-11). Turn back the clock and swap WW II stories over dinner in a stagecoach stop restaurant that went back to 1837.

We fly the converted Cessna early the next morning and head toward Iowa. Low clouds in patches; follow Flight Watch; get on top of the broken layer and bore small holes in the sky. Soon the weather improves and we pick Bloomington for fuel and lunch. No such luck—the restaurant is closed. So what the heck, it's only 20 minutes on to Peoria where the airline terminal has uninspired food and grudging service, but much better than no lunch at all. You can't win 'em all.

Ottumwa, IA, is the home of the Antique Aircraft Association. The strip was reported as 2000 feet, the day was hot, we had almost full fuel, and there was a full crop of tall corn off the far end of the field. Sure, we probably could have "hacked it," but there's always another day.

On to Des Moines where our daughter worked. Take her girlfriend for a quick ride around the field. Surprisingly, no glitches with either Des Moines Tower or Approach; but there have been other days.

To Kansas City alone where Julie is visiting friends. Another hop around town with a full load to repay them for hospitality. It's a hot evening at Johnson Executive and, with full tanks, we roll for a while before breaking ground. There's no constant speed prop on this ship.

Then, a leisurely start the next morning. Again, back in the right seat, and we have lunch at Henderson, KS. It's a good stop

Fig. 2-11. Something old, something new—a converted Cessna "151" modification puts a tailwheel back on the Cessna 150. The modification provided a magazine article during a Grumman delivery flight since London, OH, is not exactly on the main airways.

with a $2.50 buffet and many members of Beech and Cessna in the hall. Then, let's get-with-it westbound. Follow the section lines and Highway 54 over Pratt, Mullinville, Meade, Plains—all in Kansas. Then on over Hooker, Guymon and Texahoma before landing at Delhart, TX. Names of towns and VFR charts are always interesting. Undoubtedly, the high school cheerleader in Hooker, OK, is a fine young lady.

Dalhart has good service, plenty of green fuel and a coffee shop with fine pie. The long weekend approaches its conclusion, so it's on to Gallup, NM. After many flights over this grand, colorful portion of the southwest, V-190 direct to Las Vegas, NM, is a new VFR route for us. A little lonesome, but colorful with a branch of the Canadian River, headwaters of the Pecos and wagon tracks remaining for all to see from pioneers on the Santa Fe Trail.

Gallup (Fig. 2-12) has a parking area near the passenger terminal with no sign. We find out after shutting down that it's for jets only, so we fuel and Julie taxies the ship into a backline spot. We talk briefly with the airport manager the next morning, and he advises that pilots don't read signs anyhow—so what's the use? He asks if we have any suggestions to let him know. We didn't think it worthwhile.

The airport sign that we do read reminds us that if the temperature is 85 -degrees the density altitude is 9600 feet, while the actual altitude is 6468 and the runway length is 6300. It is hotter than that, so we clear the departure numbers by 100 feet at 100 knots.

Down Highway 40 to admire the gaudy signs for "authentic" Indian trading posts, truck stops and highway construction. At one spot, a truck semi has jack-knifed and stopped traffic on one side. The cab lies on its side with fuel all over the highway, but no evidence of fire.

Over Winslow we watch the clouds build up. We talk with Prescott FSS on the remote outlet and they ask for an update on what we find in the next half hour. So we'll be first into this weather, but it isn't bad. Come across the ridge, call in over Sedona and land for fuel and lunch at Prescott.

Here, a new flight training program by Embry Riddle has eight Tigers, identical to ours, chopping up the patterns. It gets a little confusing, so Julie circles twice waiting for this "mish-mosh" to subside, and we land. We park with the identically-painted Embry Riddle Tigers and almost get a free tank of fuel. The college training project is then only a few days old, and there's a little learning process. We talk with the FSS Chief and an Embry Riddle

Fig. 2-12. The Gallup, NM, airport is ringed with an unseasonal light snow. The runway and taxiways stand out sharply as the sun sets after a full day of flying.

pilot before filing and heading for Van Nuys where our Tiger will have its cage until sold.

En-route weather is interesting (Fig. 2-13). Broken cumulus are rare over the Colorado River, so we admire. Flight Watch still says that the LA Basin is socked in, so we file IFR over Daggett and go to LAX Approach for a transponder code and altitude assignment. They hand us off to Edwards Air Force Base Approach, and we're promptly vectored through a fairly energetic cumulus. As we punch out the other side, we comment to "Eddie Approach" that, "It was a little bumpy in there." He really doesn't care.

On over Palmdale with all the charts and approach plates studied. No way to eat off those "plates." We start an approach to Van Nuys and are handed off from LA Center to Burbank Approach. No, we don't have DME or a second receiver, so the controller says that we can't shoot the localizer approach. Do we have an LDA-C (localizer) plate? Yes. Okay, descend to 7000 and take a heading of 190 degrees. As we pass 7500, someone else calls at 7000 and Approach hastily requests that we climb back to 8000. We do.

Soon, there's a fairly good-sized break in the clouds and we recognize Saugus below. Rather than argue with Approach and the problems of high-density traffic on these new gauges, we cancel IFR, let down leisurely over Saugus, pick our way up the freeway through the Newhall Pass and call Van Nuys Tower. "Straight in on No. 16R. Call over the Simi Freeway."

We land, remember to ask ground control to cancel our flight plan and deliver the new Tiger to Paul Slice of Alpha Aviation. On the last leg, we've figured our expense account on the back of an American Airlines sick-sack. All the receipts and his credit card are there. Fill out the ferry pilot's squawk list on the new bird. The oil access door hinge needs tightening, the carb heat cable should be lubed, and there's a slight squeak in the right main wheel. Not bad at all for 28 hours in a brand-new airplane.

Clean the smog off the windshield, climb back into the car and battle the freeway traffic back across town. That short trip by car sure makes the long weekend by "little" airplane seem just that much more fun.

Safety in Numbers

It is essentially safer and usually more fun to deliver two aircraft at the same time. I've been able to arrange this on a number of deliveries out of Wichita, KS, and Albany, GA, but primarily from Osceola, WI. While the safety-in-numbers routine

Fig. 2-13. En route weather is always challenging. Western mountains reach for the base of the overcast and the ferry pilot thinks about changing his destination airport.

helps in case of any mechanical or weather problems, and makes it possible to complete the delivery if one avionics package decides to quit, it also has the advantage of providing ready-made company on the ground. Solo cross-country is a great experience, but some of these smaller towns—unheard of by the pilot until it was time to quit flying on that particular day—do tend to get a little lonesome.

There are other advantages in the multiple airplane ferry. If two new ships have different destinations, considerable travel expense can be saved at off-airline destinations because one aircraft can be delivered en route and both pilots continue back to their home base in the other ship at no additional expense.

This two-plane delivery worked out nicely with a Citabria and a Decathlon headed for the West Coast. The Decathlon went to Phoenix with the Citabria destined for Santa Paula. As usual, the trip came up on short notice with the second airplane a last-minute addition. Genial Ed Lamb, then Western Representative for Bellanca, asked for the two ships and requested that I get a second pilot. As it happened, my daughter, Dana, had time available and was current in the Citabria. So we headed for Minneapolis (Fig. 2-14).

Citabria's Chief Pilot, Dick Johnson, picked the two of us up at MPS in a Bellanca, and we headed for the Osceola, WI, factory. Only one airplane, the 150-hp Citabria, was at the Wisconsin factory. The Decathlon was in Alexandria, MN, where the speedy four-place Bellancas are built. Dana and I loaded our gear in the Citabria and I stretched out in the back seat while she bored through the head winds to Alexandria.

Dana had never before flown the Decathlon with its constant speed prop, so we traded airplanes and took off for Redwood Falls, MN. The following day we bucked heavy head winds as far as Topeka, KS, and landed on the parking ramp there to eliminate any crosswind taxiing. Again it was windy, and we had line personnel "wing-walk" the light taildraggers into takeoff position—again on a nearby taxiway. Under these conditions, ground roll is something under 100 feet.

The wind slackened at Dalhart, TX, and it was time to give Dana five landings to make her legal in Decathlon. With the controllable prop and injected engine the only basic differences, the checkout was routine. Then she took the gaudily-painted aerobatic ship to Tucumcari, NM, and then to Albuquerque.

When the Decathlon was safely delivered at Falcon Field near Phoenix, Dana went up for an hour of aerobatic dual with a

Fig. 2-14. Dana Downie relaxes in the sky during a two-plane delivery to the West Coast. The sunburst paint job on the Citabria has always been popular.

local instructor. On landing, both brakes were locked and the ship ground-looped without damage. Subsequent inspection showed a mechanical malfunction in the parking brake installation that locked the brakes the first time the airplane was slow-rolled. A telephone call back to Dick Johnson in Osceola led to the discovery that every Decathlon on the line at that time had the same problem.

Somewhat wiser and relieved that no injury or damage had occurred, Dana and I went on to Santa Paula where the Citabria was delivered.

Two Scouts West

Two of the first Citabria Scouts flown to California were picked up by Bob Weston and myself at the factory. Weston is a full-time plumber (so our house has no leaky faucets) and an avid weekend pilot who can fly taildraggers. While he has only slightly more than 400 hours, wild weather doesn't make him particularly nervous and an occasional ferry trip is great for his morale (Fig 2-15).

Therefore, we rolled off Western's midnight "red eye" at Minneapolis and headed for Holman Field in downtown St. Paul. This particular dual delivery was somewhat different than most because Weston had never before flown the Scout, and I had the go/no-go responsibility for checking him out in the second ship or bringing him back with me. The following material was used by AOPA Pilot Magazine.

"Check him out when you get back to St. Paul," cautioned Mike Dewey, the Citabria dealer in Santa Paula, CA. "If he's okay, bring two ships. If not, just bring one."

So we climbed into Scout N57669, one of the prototypes and started out from Holman Field in downtown St. Paul. The weather was drizzly and just above VFR minimums. N57669 had radio problems, so we decided to take it back to the Osceola factory, some 40 miles northeast, to have the unit changed. We shot a series of practice landings along the way, and Weston proved that the Scout is, if anything, easier to fly than a stock Citabria.

With new ferry pilot Weston completely legal, we went into the factory. It wook Test Pilot Dick Johnson less than an hour to procure a new Narco Escort 110, install it, check-fly the set for both nav and com, complete the paper work and have us ready to go. However, the weather wasn't good, even by California fog-'n-smog standards, and the guessers were forecasting it to get worse. When a Citabria delivery was set for St. Paul, I decided to catch a ride and bring back our No. 2 airplane, N86721. Therefore, we'd have both planes and pilots at the same starting point the following morning (Fig. 2-16).

This remain-over-night decision was encouraged, in part, by the hospitality of this small town, population 1000. Former Citabria plant manager, Rollie Seager, had just completed the St. Croix Valley Inn within a half-mile of the Osceola Airport. Rooms, meals and a bar make a complete pilot's stop and transportation to the airport is always available. We watched the rain come down in solid sheets until late in the evening.

Our morning weather check with the Twin Cities FSS was just the beginning of things to come. It was still wet and windy, but continued VFR was reported to the West. We picked Willmar, MN, some 120 miles to the west and taxied out gingerly.

Fig. 2-15. Plumber-pilot Bob Weston flies an early Citabria Scout after a check-out near the factory. This ship eventually became a tow plane at El Mirage Soaring Center.

The "Ops Manual" for the Scout lists a demonstrated crosswind component of only 15 knots, although I suspect that it will handle considerably stronger crosswinds. Therefore, our choice of landing spots was dictated by into-the-wind runways. In this respect, AOPA's ever-ready Airport Directory was essential.

A quick circle of the factory indicated there was no way for VFR westbound. The only bright spot on the horizon was to the south—so that's the way we went. We bounced in gusty air around the Minneapolis and St. Paul control zones and stayed with the broad freeway to Mason City, IA.

The Mason City FSS gave surface winds of 22 knots, gusting to 35 from 270 degrees. The broad runways there are 30 and 35 with almost all the other airports in the area heading north and south.

Not wanting to drag a wingtip 1500 miles from home (or anywhere else), we circled the field and elected to land across the runway, a short taxiway and runup pad at the far SE corner of the field. With a full-flap (30-degree) stalling speed of 47 mph at full gross weight, our touchdown was incredibly short.

In response to a radio request, the gas man and the FSS Chief—on his lunch break—walked our wings in. Great people!

As with any new production airplane, we had a few interesting developments. Both Weston and I found the front seats too low and we sat on flight jackets and topcoats all the way west. Even with this extra padding to lift our eye-level, there was still ample room between the top of the head and the cabin roof. I'm a few inches taller than Weston and reversed the front seat cushion to give me an extra two inches of leg room.

The only potentially difficult situation on the entire trip occurred shortly after takeoff from Russell, KS. When I pulled in to take a couple of color slides, I could see an intermittent surging stream of green 100 octane coming off the top of Weston's right wing (Fig. 2-17). Closer inspection indicated that the gas cap was on, but every time that Weston hit a bump, he spewed fuel all over the sky. Surprisingly, it was his left fuel gauge that started to go down, while the right tank showed fairly full. When we landed after a one-hour flight at Dodge City, Weston took 22.6 gallons—most of it in his left tank—while my ship used 9.8. A faulty alignment of clips on the fuel caps made it impossible to get them screwed-down a full 90-degree turn without using a pipe wrench.

Reference paint marks should be placed both on the top of the wing and on the gas caps to assure they are fully locked.

Fig. 2-16. Simple instrument panel on Citabria Scout as seen in flight. Single radio is an Escort 110. No blind flight instructions were to be expected.

One of the more interesting aspects of this ferry flight was our choice of into-the-wind takeoff and landing pads. Of 13 landings on the ferry flight, we used conventional runways seven times. At Dodge City, with winds at 22 knots gusting to 30, we took off straight into the wind from the tiedown ramp. Later that day at Tucumcari, NM, we had similar winds and advised the FSS, "At pilot's discretion, we'll use the small (+ - 600 feet) taxiway leading directly to the tiedown area." Parking ramps at Dalhart and Daggett put us directly at the gas pits.

We had our up-wind wing walked to the remaining portion of runway 26 at Tucumcari and rolled only a few feet before breaking ground. Soon a pick-up truck towing a horse trailer was passing us on the highway below—and he was going uphill! We ran a quick ground speed check that showed 70 mph. We had planned a landing at the Coronado Airport in Albuquerque, since I doubted that the joint-use International Airport would have appreciated a request for a flight of two to land in the parking area of the Cutter Flying Service.

But we never got to Albuquerque. As we plodded along into the wind, Weston and I discussed our probable fuel situation on landing, using the plane-to-plane 122.9 frequency. We'd have enough, but not too much. About that time, the pilot of a Cessna T-41-B—a surplus USAF primary trainer similar to the Riems Rocket—called in to advise that 100 octane was available at Moriarty, just south of the Otto VOR. The Cessna pilot turned out to be Lt. Bill Eddleman of the New Mexico State Police Aero Detail. He offered to check with his low-frequency police-band radio to assure that operator Paul D. Scates was at the airport. He confirmed that the FBO was there to help us with tiedown and fuel, so we made our first visit to Moriarty, NM. Lt. Eddleman landed ahead of us and we tied the Scouts carefully before riding into town for lunch in a black-and-white squad car driven by Tri-Pacer owner Sgt. Finis Roberts. More great people!

After lunch, a recheck of the winds sounded as though we'd be thrashing around in the bumps and getting nowhere. Lt. Eddleman was headed for Santa Fe where the wind was right down Runway 28, so the two policemen and the FBO walked us to the runway where we hopped into the air in no time at all (Figs. 2-18 and 2-19).

We relaxed around a windy motel pool that afternoon and left a wake-up call for well before daylight. On flight plan, we enjoyed the colorful solitude of the Edolleta Mountains as the shadows shortened and the wind began to pick up. After breakfast in Gallup,

Fig. 2-17. Formation flying like this for photos showed fuel leaking from a gas cap on Bob Weston's new Scout. A faulty fuel cap was fixed after landing at Dodge City, KS.

Fig. 2-18. Bob Weston taxies his Citabria Scout in a high wind on the ground at Moriarty, NM. A State Highway Patrolman holds down the up-wind wing.

Fig. 2-19. Two members of the New Mexico State Police takes us to lunch at Moriarty, NM. State Pilot Lt. Bill Eddleman is at the left; Bob Weston, a State Patrolman and the author await fueling before leaving the planes.

we plodded into Flagstaff for the best weather briefing of the entire trip.

At Havasu City, we ate and debated about calling it a day. However, coastal fog was forecast for Santa Paula the following morning, so we strapped on the two Scouts and headed toward the sunset. There was a just-to-be-sure fuel stop at Daggett and a dusky landing at Santa Paula. After 8½ hours in the air, most of it in turbulence, we crawled out of the two new ships feeling that we'd put in a full day (Fig. 2-20).

"There was just one thing I didn't like on this whole trip," said Weston as he tied his ship down in Santa Paula. "That was having to give up the keys."

Ten Ladybirds

A four-day trip of ten Cessna 150's from Wichita, KS, to Van Nuys, CA, provides some interesting experiences in the logistics of mass movement. The following "Flight of Ten Ladybirds" was published in "AOPA Pilot Magazine."

There wasn't much different about the 10 new Cessna 150's that took off from the company's delivery center at Strother Field, Wichita, KS. Of course, all the pilots were women and members of The Ninety-Nines Inc. The first eight planes headed west into

24-knot head winds without delay. One remained frozen to the taxiway for a few minutes until ice was removed. The other waited for a flap retraction circuit to be replaced. It was just after dawn on a Sunday morning and there was a short delay in waking up one of the nearby mechanics. Perhaps the only unusual item on this 10-plane delivery flight was the inclusion of a male reporter-photographer team (Figs. 2-21 and 2-22).

It all began with a sweet feminine voice on the other end of the phone. "We're bringing a flight of 10 Model 150's back from the factory. How would you like to come along and do a story on the trip?"

I said I'd call her back, and checked with "my gal," also a Ninety-Niner, who displayed a tolerant grin and commented, "I'm sure that you'll find the trip interesting."

So that's how I happened to be sharing the airways with a competent group of "Ladybirds" from the San Fernando Valley Chapter of the Ninety-Niners on a four-day trip to California. Yes, four days! Did you ever hear of rain in El Paso, TX, and zero-zero conditions at Deming, NM? Do they ever have late afternoon cloud bursts near Cochise, NM, and rain falling from a low overcast in the Los Angeles Basin? Sure they do, but not usually all on the same flight.

The 10 Ladybirds all dressed the same every day. Each had been presented with a skirt, two blouses and a "shell." I didn't

Fig. 2-20. Santa Paula Airport as seen on final approach during an unusual landing on Runway Four to the northeast during "Santa Ana" wind conditions. Both new Scouts were delivered here.

Fig. 2-21. Ten "Lady Birds" picked up ten Cessna 150's for delivery to Van Nuys, CA. The author went along to do a magazine report on this delivery flight.

know it, but that's a sleeveless sweater that will drip-dry at overnight stops. The "berry red" outfits were a gift from Bill Pattison, the director of Bardahl Aviation Products, who jetted to Wichita with us to make sure that a can of the company's newly-approved aviation lubricant was aboard and a prominent decal placed on each aircraft.

Flight experience of the ferry pilots ranged from two to 25 years and from 190 to 3600 hours. All but two had at least a commercial license. Three were active flight instructors. A "buddy" system was used to team a high-time and low-time pilot, who shared rooms each evening to expedite flight planning for the next day. No one was ever really lost and what confusion did occur was caused by 10 airplanes all arriving at the same airport at the same time.

Five members of the group had been on a similar safari of 12 aircraft for Van Nuys Skyways in February. Both deliveries were used as "brush-ups" for the Powder Puff Derby. (See Figs. 2-23, 2-24 and 2-25).

That's the background to begin the saga in which I found myself folded up in the just-big-enough right seat of a 150 with Mrs. Libby Svenson, a mother of five, with 190 hours in her logbook. By the time we reached California, she'd added almost 20 hours more.

Fig. 2-22. Cessna Delivery Center at Strother, KS, where 150's and 172's are completed and flown away (Courtesy Cessna Aircraft).

Fig. 2-23. '99'er Bettye Fry, in mink coat, checks gas caps before taking off from Strother, KS, in a Cessna 150.

The 150's were spread out in pairs at dawn over the flatness that is Kansas, chatting easily on a busy 122.9 plane-to-plane channel. Since the 150's were to be used on a GI commercial and instrument flight training program, each had a Narco Mark 12A, complete with glide-slope receiver and three-light marker.

The original "sweet voice behind the telephone" turned out to be Ardie Trenholm, assistant flight operations manager for Van Nuys Skyways. She was "mother hen" on the ground, taking care of the details of housing, transportation, publicity, box lunches and the bills.

"Mother hen" in flight was Lola (Mrs. Eduardo) Ricci, wartime ferry pilot with more than 2400 hours spread over 25 years of flying. She called the shots for go/no-go after checking weather along the route and did a fine job considering the variety—all bad—that we encountered.

After leaving Strother Field, everyone settled back to see just how bad the head winds would be—and they were bad. Our ground speed averaged 80 mph most of the time. Lola had planned our first fuel stop at the Perry Lefors Airport in Pampa, TX, just 245 statute miles upwind. However, some of us didn't make it.

Fig. 2-24. Cessna 150's are lined up for a dawn takeoff by ten "Lady Birds" at the Strother, KS, Delivery Center.

The trainer version of the 150 has 26 gallons of fuel, but only 22½ are listed as usable. The remaining 3½ gallons settle to the low trailing-edge corner of the tank where the fuel sump is located. The fuel selector valve is either "on" or "off" with no left or right; so there's no chance of coaxing this unusable fuel from the tanks.

As the little fuel gauges began flirting with the big "E," Libby Svenson and I held a little flight conference. We looked in the 150's operation handbook and learned that the 100-hp Continental engine can burn as little as 5.1 gallons an hour at 2500 rpm and 2500 feet, leaned out. However, the 10 little birds were flying right on the deck to stay out of even worse head winds.

We had been in the air 2:45 when we reached the Hemphill County Airport, located at the town of Canadian, TX. By mutual consent, we decided we didn't want to sweat out the remaining 58 miles to Pampa and advised "airborne mother hen" that we were diverting to Canadian for fuel. By the time we were over the airport, three other pilots had reached the same decision.

The latest charts advised that Hemphill County was "attended 24 hours," but it was Sunday morning and no one was around. I walked across the airport to the nearest house for a telephone.

We were able to get Texaco dealer Bob Ward out of church to fill the four ships (Fig. 2-26). One took 19.8 gallons; we took 18.5. Our fuel consumption on this initial leg figured at 7 1¼ gallons an hour, which would have landed us (hopefully) at Pampa with a half a gallon of usable fuel. And, as a long-time professional airborne coward, that's cutting it far too close. Of the six 150's that went on to Pampa, one took 22 gallons and two took more than 21. Right about then, the optional 36-gajlon tanks ($195 exchange) would have been worth the price.

It was with a clear, cowardly conscience that the four of us departed the friendly dirt airport at Canadian with a planned fuel stop at Clovis, NM. However, our winds were right on the nose, and it didn't take long to show another 2:45 to three-hour flight in store. We landed at Amarillo to top off and filed "via the highway" to Roswell, NM.

We headed into Roswell, where I had landed before at the old Roswell Municipal Airport. The latest available Sectional charts

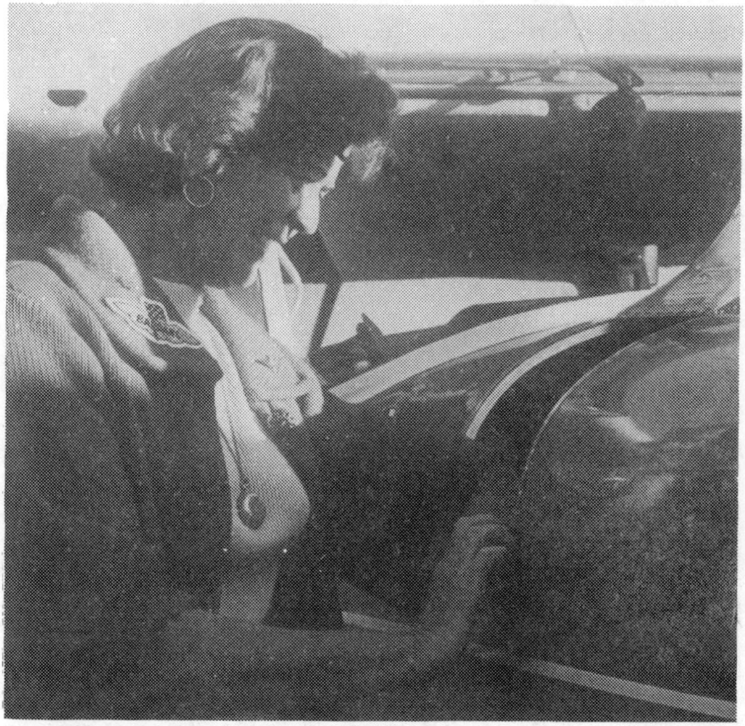

Fig. 2-25. "Mother hen" Lola Ricci checks the oil on her Cessna 150 during a fuel stop.

Fig. 2-26. Several Cessna 150's make an unplanned fuel stop at the Hemphill County Airport, Canadian, TX. The group asked the fuel dealer to come out of church to fill the trainers.

and Jeppesen inserts showed the former Strategic Air Command base at Walker Field was closed. Perhaps a hot-off-the-press Airman's Information Manual had something different, but Mrs. Svenson and I went blithely on our way, with other Ladybirds tagging along. We called "Roswell Municipal" on the appropriate 118.5 frequency and were cleared for a "straight-in" on Runway 17. Somewhere in the tower discussion, we heard the phrase "Roswell Industrial Air Park," but it didn't seem to apply.

We came in on a conservative high final approach and could see the big, white 17 on the end of the runway. We were within a mile of touchdown before I began to become suspicious about the lack of activity at the airport. There were only two airplanes visible on the ramp and there should have been at least four of our flight alone on the ground.

"Let's go around and take another look," I suggested, and Ladybird Svenson poured on the coal. We pulled up and called the tower again.

"No," they advised, "the Roswell Industrial Center Air Park is five miles south of town. You are cleared to land straight in on Runway 17, that's the one just to the west of the large black hangars."

We had almost gone into the wrong airport. Later I asked the tower operator what had happened. Just six weeks before our visit, Walker Field had been opened by the City of Roswell and the other airport closed. I asked why the "new field" wasn't called Walker-Roswell to eliminate confusion and was advised that, "We're not permitted to do that."

So everyone had their own airborne guessing game. At least until new charts are available, it would seem practical to permit the use of "Walker-Roswell" rather than "Air Park."

We fueled, checked weather and the time of darkness for El Paso, just 160 lonesome miles distant. Our flight of two planes had now grown to three. We made sure that everyone had a flashlight aboard. Then, with the comment, "Let's keep fairly close together and check in by radio every 10 minutes," we took off for the Pinon omni and El Paso.

If we did not hit Pinon on the nose, we planned to return to Roswell. There were king-sized rain squalls and associated lightning to our right over Alamogordo and the Holloman test range. Rain was also visible over Carlsbad to the left. Fortunately, the area over Pinon looked respectably clear. So, with navigation lights on and beacons flashing, the three little 150's headed out across the high Southwestern desert.

By the time we reached Pinon, I had a stiff neck from turning around to count the running lights through the omni-vision rear window.

We called El Paso approach control as "a flight of three" midway inbound from Pinon omni. I was a bit surprised when the operator asked for an identifying turn from 207-to 180-degrees. I wondered how many "flights of three" he might have had on his scope. However, El Paso's approach control was excellent and we were back on course within 90 seconds with vectors to keep us away from departing airline traffic.

I finally found out why Cessna put those rear-view mirrors in their cockpits. Every gal I flew with, and others that I asked, used the mirror to freshen up her lipstick before starting her landing check list.

When we changed over to the tower frequency, "we were cleared to land on Runway 35. The tower asked, "Do you plan to land as a flight of three?"

It had been a long day, with over 8½ hours in the air, so I couldn't resist the temptation to pick up the mike and reply, "Who knows?"

All was quiet and we were cleared to land. The other two 150's followed closely on our tailfeathers and it was only over dinner that I learned that pilot Virginia Rainwater, who had 240 hours with a commercial certificate, had so much grease on her windshield that she had to land looking out the side windows. A new prop seal was put on her 150 the next morning.

The last three of our little birds, led by "ground-bound mother hen" Trenholm (Fig. 2-27), came in behind us just a little after dark. Thirty minutes later it was raining Texas "cats and dogs."

A tour of Ciudad Juarez, Mexico, had been planned for the evening, but everyone admitted that it had been a long day and, to my knowledge, no one crossed the border that night. "Airborne mother hen" Ricci promised to check weather at 5:30 and call each room if an early takeoff were planned. I slept until 10 a.m.

It was well after noon when the weather showed any promise, and we took off as a flight of 10 at 30-second intervals. Because of the limited range of the 22½-gallon "eye-droppers," we

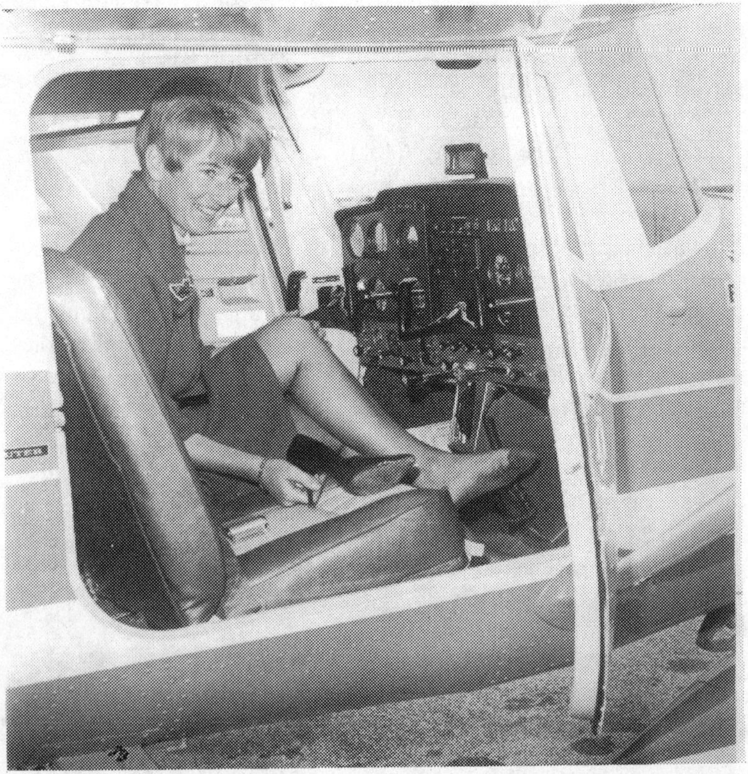

Fig. 2-27. Ardie Trenholm wears high heels on the ground but flies a Cessna 150 in her stocking feet.

flew only 100 miles to Deming and topped off our tanks before continuing through some interesting weather to Tucson, AZ.

I flew this leg with Mrs. Patricia Pitcock who had 250 hours and a private license. We spent part of the trip comparing notes about her 13-and 16-year-old children and my somewhat older daughter. One of the things that she always carries with her maps is hair spray.

There's an FSS at Deming, but no tower. Therefore, all 10 radios remained on 122.9 for landing. With such a short hop, everyone arrived at the same time and the traffic was terrific. After being cut out twice, we landed and topped off.

Next was the two-hour flight to Tucson—maybe! Weather was spotty and "mother hen Ricci" prudently advised Cochise County was an alternate airport. It was interesting to listen to the air-to-air frequency as the 10 Ladybirds discussed airway intersections versus the pipeline or the "iron compass." As the scattered rainstorms became broken, more and more of the teams of "buddies" detoured the weather and picked up either the railroad or the highway.

Bettye Fry—commercial rating and over 800 hours—flew as our "buddy," remaining perhaps 500 feet behind for the entire trip. We discarded omni for the "iron compass" and turned the corner near Benson, AZ, just as the sun was setting. Even though the two of us were the last to take off from Deming—Bettye stayed to sign the gas chits—we all came into Tucson at the same time.

It was a rather interesting 15 minutes. Approach control advised "the flight of 10" to call in 10 miles out for landing on Runway 30R. The tower frequency became somewhat congested and the 10-mile reporting point came down to five and then three miles before we could get our number in. Finally, Mrs. Pitcock turned on a long, high final approach while I kicked on a landing light to make us a little more conspicuous.

The Cessnas all had numbers from 36S to 60S, all ending in "Sierra," and the Tucson tower operators had their problems. As Mrs. Pitcock and I were nearly halfway down the 11,500-foot runway, still at 1000 feet and showing a landing light, we'd about decided to go around when the tower advised (as nearly as I can remember it): "Cessna nearing mid-field with a landing light showing, cleared to land. Cessna crossing the approach end of 30R with a light on, cleared to land. Cessna following the aircraft crossing the threshhold cleared to land. All other aircraft pull up and circle to the right."

Fig. 2-28. Lady Birds discuss the days adventures at the Tucson International Airport after sunset. The ten trainers gave the tower a slight problem.

Right in the middle of this "Chinese Fire Drill," approach control handed off an American Airlines 707 for a straight-in approach to Runway 12, which put the big jet nose-to-nose with one of the landing Cessnas.

"I never turned off a runway so fast in my life," said the unidentified Ladybird later.

When I talked with Tucson Tower on the intercom a little later, he said that the flight of 10 had been "a little hectic."

As we waited on the ramp for all the 150's to be tied down, we saw a most unusual sight to the west in the last of the setting sun (Fig. 2-28). It looked for all the world like a UFO come to life, complete with bright light and rapidly-expanding vapor tail. FAA personnel received many queries, but it wasn't until the next day that we found out we had watched the climb of a satellite launched from Vandenberg Air Force Base, more than 650 miles to the west!

A press and TV arrival party was scheduled at Van Nuys Skyways for 2 p.m. the following day, so it was decided to get up at 4 a.m. and make a pre-dawn takeoff to assure plenty of time for arrival.

Everyone was at the airport, perhaps not bright-eyed, long before dawn. "Airborne mother hen" Ricci called the shots and it was decided that I rejoin Mrs. Svenson for the first leg to Phoenix. She had never made a full-night takeoff or landing, so I rode "shotgun." She did very nicely. The 150's were all well covered with moisture, but not frost. The windshield defroster vent on this model is a take-off from the cabin heater, and it's not too effective. By the time we had taxiied half the distance of the two-mile runway, there was a little clear hole that was just big enough to see through for takeoff.

The sun came up as we approached Phoenix, where everyone landed for fuel and coffee (Fig. 2-29). Next stop, Thermal, CA. This time I co-piloted for Bettye Fry so that she could get a little shut-eye and conquer a not-enough-sleep headache.

The 10 little birds arrived at Thermal with what we thought was plenty of time to spare. Bardahl's Bill Pattison again met the group, this time with his gaily-painted Cessna 182. Bardahl had just been approved for regular light aircraft use and this was the company's first promotional effort in general aviation.

Since it would make a better—and far less complex—story for reporters and the TV stations if photographer Jim Zonlick and I weren't aboard, we joined Pattison for the final leg to Van Nuys.

At Van Nuys, 2 p.m. came and no one showed up. Finally "mother hen" Trenholm called in from Palm Springs. Weather in Banning Pass wasn't good and ceilings were lowering all the way to Van Nuys. After a telephone conference with Van Nuys Skyways general manager, Allen E. Paulson, it was decided to round

Fig. 2-29. Cessna 150's on the ramp at Phoenix for fuel and coffee.

up all the IFR-rated pilots available, send two planeloads to Palm Springs and get everyone home that night.

In a way, it was a shame not to delay the arrival until things were VFR, but who can outguess the weather? We loaded Cessna 310 and the Bardahl 182 for the flight to Palm Springs. While pilots were mulling the problem over in Pacific Airmotive's ready room at Palm Springs, Audrey Schutte called, "Hey Downie, let's go!"

We did, since it was getting darker—and wetter—at Van Nuys every minute. Mrs. Schutte had two teenagers, worked full-time as a flight instructor at Van Nuys and had logged 3600 hours in 11 years. Her husband was on a ferry flight at the same time, bringing a National Guard helicopter from Ft. Benning, GA.

The first half of our trip was not uncomfortable, if you like blowing sand and moderate turbulence. Then it began to get dark and dewey, so we picked up the San Bernardino Freeway inbound and dutifully called Ontario International's tower for clearance through their area. Then it was a quick change to Brackett Tower for an okay over their brightly-lighted runway.

We began hearing new voices on the 122.9 frequency. It turned out to be both police and commercial "airwatch" aircraft reporting traffic conditions to the ground-bound freeway fighters. Since our weather was deteriorating rapidly, Mrs. Schutte called one of the traffic control choppers to advise that a flight of 10 Cessna 150's would be passing through their patrol area. From then on in, no less than three traffic control aircraft came in with up-to-the-second weather reports. One even attempted an air-to-air interview with "mother hen" Ricci.

The endless string of red taillights made a fascinating pattern as we "drove down the freeway" at something over 1000 feet and below the cloud deck. Mrs. Schutte and I compared freeways as we navigated by road map to Van Nuys. She called the tower, made a prudent close-in pattern, chopped the power and turned off at the first high-speed taxiway. The other 150's followed in rapid order, with the exception of two that diverted to Santa Monica where the weather was a bit better. Later in the evening, all 10 little birds were safely parked outside the brightly-lit Skyways hangar.

Yes, it had been an interesting trip.

Chapter 3

Getting Down To Business

Let's take a look at some professional ferry services and meet three well-known ferry pilots and one well-known personality who once ferried a plane.

It's a fine line between the friendly family FBO and the full professional ferry service. The larger the flying service, usually the more impersonal. But each of the three businesses we've interviewed here have a very personal interest in each of their pilots and his or her capabilities.

How Pipers Come West

ATCO, the Alaska Transportation Co., of San Jose, CA, has handled western deliveries of Piper aircraft for many years. The "Alaska" name is from a corporate structure of a previous company and the majority of ATCO's deliveries are in the western contiguous states.

Marge Maus heads up ATCO's ferrying division. I have ferried two Cherokees back for her, both made after a press trip directly to the Vero Beach factory aboard the Los Angeles Dodgers' Boeing 707. The connections are just about perfect since the Dodger spring training camp is within a couple of blocks of the Piper Delivery Center (Fig. 3-1).

ATCO's pilot qualifications are simple. Marge Maus explains that everyone who ferries their aircraft must have a commercial license and 1000 hours. Many of their pilots come from nearby

Fig. 3-1. Going and coming—Los Angeles Dodgers' Boeing 707, with Captain Lewis Carlisle in the left seat, served as a way east to Vero Beach on two delivery flights. In contrast, the Cherokee 140 ferried west does look a little smal .

Fig. 3-2. Piper Cherokee assembly line at Vero Beach. These aircraft were completely assembled and test-flown before going to the paint shop.

colleges with aeronautics programs and are usually flight instructors working on their ATR's. Except for a minor "fender bender" on occasion, ATCO hasn't lost an airplane in a 13-year period, and they average 250 long cross-country deliveries a year (Fig. 3-2).

A sheet with the following instructions is given each ATCO pilot:

Delivery Pilot Instructions

Factory Pick-up—Prior to signing the factory acceptance card, inspect the aircraft condition, that all optional equipment listed is installed, that all loose equipment (aircraft paperwork, headset, mike, lighter, tow bar, etc.) is accounted for, and all fuel tanks are full.

Equipment operation – Prior to departure, check ground operation of all equipment possible. Have any discrepancies noted corrected.

As soon as possible after airborne, check operation of all equipment. If any discrepancies are noted, return to factory and have them corrected.

En route – Conduct all operations in accordance with current FAR's.

- Single-engine aircraft can be flown day VFR only;
- Multi-engine aircraft can be flown at night only by IFR-rated pilots;
- Multi-engine aircraft can be flown IFR only by pilots specifically approved for IFR flight;
- Aircraft are not to be flown in moderate or heavy rain or icing conditions;
- Aircraft operation is to be as specified in the aircraft manual;
- Each flight must be planned to provide a 25 percent fuel reserve until average fuel consumption can be established;
- Subsequent planning must Provide either a 45 minute or 10 percent of total flight time reserve, whichever is greater.

Should repairs be required, try to have work done by a Piper dealer (Warranty Station if avionics) under warranty. If any charges are made for the parts, return defective parts to ATCO. If charges exceed $50.00, phone ATCO for approval.

Aircraft must be fueled and tied down at overnight stops. If the possibility of bad weather exists, aircraft is to be hangared if at all possible. Passengers are not permitted unless prior approval is obtained from ATCO.

Delivery – Complete ATCO Delivery Flight Report. If delivery is to other than ATCO location, have report signed by person you deliver aircraft to, and mail the report to ATCO. If delivered to ATCO office or aircraft warehouse location, turn report over to person accepting airplane.

Area ferry – When aircraft is ferried from other than a factory location, all information applies, except factory acceptance card signing.

Ferry pilots are also requested to fill out a detailed "squawk sheet" on their new aircraft. This detail is requested by the owner on all ferry flights and should be turned in at the same time as the expense account. ATCO uses a regular Piper customer "first impression survey," listing the aircraft model, serial number and registration number. The form has blanks to fill out covering general appearance, interior and exterior, and electronics—including avionics and auto pilot. Aircraft systems are itemized for powerplant, electrical, controls, instruments, hydraulic, fuel system and environmental, which covers heating, air conditioning and pressurization. The Piper factory wants to know about the aircraft preparation and requests comments on factory delivery personnel (Fig. 3-3).

In reply to one of these queries, Mark Heath, an ATCO pilot who is an engineer for United Airlines, gave the following detailed report:

Re: Delivery of PA38-112
N2465K
Above aircraft delivered to Flood Aviation SQL. Keys left with aircraft in usual location.
Tach time: SQL 29.45
MDT 3.50
En route: 25.95
Mechanical: Note: Customer Response Form inadvertently left in aircraft.

The oil temp ran high and the oil pressure ran low. This was affected to some degree by outside air temp. At 10,500 ft OAT +5C the temp was just to the left of mid position of the upper green band (there are three divisions of the green band). At this altitude the pressure was in the white (covered it from view). At low altitude and high OAT the oil temp indicated approx one needle width from

Fig. 3-3. Piper Cherokee 235 over the Arches National Monument in Utah. Ferry flying gives a pilot a chance to explore areas of the country—as long as he doesn't overdo it!

the red and the pressure indicated just at the lower edge of the white (high edge of the yellow). Oil quantity had no effect (adding one qt. did not change things).

The flap handle latch failed to 'lock' unless air speed was very slow (well below the high end of the whitearc). It tests ok on the ground but not ok airborne. This needs some real looking into.

Re the left door top latch—apparently, the rectangular loop (on top of door) that the latch pulls in had broken and had been welded. A burr or tip of welding remains on the top-outboard corner of the loop. When opening the door, this burr catches onto the door seal that mounts onto the cabin side of the door opening. This pulls the seal out when door is opened.

Above three writeups left with aircraft.

Just for information, the other aircraft (N2444K) had the following problems:

- VHF transceiver loop.
- Flap lever lock button was stuck in, which would never permit use of flaps.
- Fuel would not drain from engine drain valve.

The Tomahawk has outstanding visibility. The fuel selecting and indicating arrangement is tops! Considering the size of engine, it performs very well.

Marge Maus remembers that this same airline Engineer filled in one post-delivery report with the comment, "The clock is five seconds off."

Fig. 3-4. Routine Piper Cherokee delivery at Vero Beach. Dana Downie stands on the wing with a silly "tourist" hat prior to loading a Cherokee 140 to deliver in California.

"Unfortunately, this is the exception rather than the rule," commented Marge Maus, referring to the amount of detail (Fig. 3-4).

Performance Aircraft Requirements

One of the larger West Coast dealers is Joe Geiger's Performance Aircraft Inc., with offices in Long Beach and Hayward, CA. There's considerable ferrying done by his pilots, almost all of whom must have a minimum of a commercial license and 500 hours. Occasionally, he will authorize a private pilot, but only one that he knows personally and with whom he has flown.

"I want a pilot who will keep 'right on truckin'," explains Geiger. "I don't want a pilot who will stop for repairs if one of two radios go out or there are a couple of thunderstorms on the horizon. Shucks, anyone who flies across the middle of this country in the summer is going to dodge thunderstorms."

"It's tough to get to be a professional ferry pilot unless you're willing, able and qualified for overseas deliveries. Too many pilots are willing to do it for expenses."

And that's what Geiger pays—Air coach fare to the factory, gas and oil, and reasonable motel and meals. He has what he calls "a pecking order" on available aircraft. His No. 1 choice is working press—trade magazine and newspaper people who have adequate qualifications. That's one reason why I've made three ferry trips for him. No. 2 on the list are his employees who need to visit the factory for a first-hand look at what takes place there. No. 3 are customers we have faith in, explained Geiger. "We don't encourage our customers to ferry out their own aircraft. We'd rather have a high-time pilot behind the controls, because if something comes 'unglued,' it will probably happen within the first five hours. Then come some of the 400 members of flying clubs maintained by High Performance. Also on the ferry list are club instructors or employees on vacation.

"We can get all the youngsters we want to ferry the light, slow trainers," said Geiger. "They'll fly anything just to build up time, up to that magic 1500-hour mark that will make them qualify quickly for an airline job.

"If we have older, used aircraft, we'll use our own salesmen who are getting paid for the job and assuming any risk that goes with a 'pre-owned' aircraft."

Frequently, the company president goes back to the factory to pick up the first of a new model or a new engine combination. Geiger recalls picking up a Maule from Moultrie, GA, and making a maximum performance climb out of the factory. "It was a damn good thing that I did climb steeply," he commented. "About eight miles out I began to have that hot metal smell. I turned back and soon I was getting only 1500 rpm at full power. As I eased back on the power, once I had the factory field within range, the engine quit cold. I tried the starter in the air and nothing would budge, so I landed 'dead stick'. Two hours later that engine was still frozen up so tightly that two of us could do chin-ups on the prop and not move it. We found out later that the rings had been installed improperly on one cylinder of the new engine."

It isn't often that a pilot will bring back the same airplane twice, and make a wheels-up landing both times—but it happened to Geiger (Fig. 3-5). He was picking up one of the first Mooneys manufactured in Kerrville, TX, that had a retractable gear system. As Geiger explained it, the original electric gear retraction system had been manufactured by Dukes and later replaced by an ITT unit. The original system cycled up and down in three seconds, while the replacement unit took about seven seconds to retract and five to come down.

After a brief sales tour, Geiger was departing the factory at night with the sleek retractable. He hit the gear switch, the gear came up and "pow!"—the gear warning horn began to blow. He

Fig. 3-5. Joe Geiger, President of High Performance Aircraft in California, inspects the landing gear of an older Mooney. He was able to land the same airplane twice gear-up—and the factory finally picked up the tab!

could hear the electric motor run as he attempted to recycle the gear, but nothing stayed either up or down. He tried to crank the gear down without results. He heard Sales Manager Don Cox and Regional Manager Jeff Conrad calling on Unicom and advised them of the predicament. They flew alongside with their landing lights on and ascertained that the gear was hanging free-down, but unlocked.

Since it would be less complex to make repairs at the factory than anywhere else, Geiger picked out the main paved runway at Kerrville, lined up, shut everything off and landed. Damage was limited to a bent prop and minor skin and flap damage. However, in bending the prop, it required that the engine be removed and the crankshaft inspected.

"If you have any choice, always land gear-up on pavement. The best is a newly-rocked runway where you'll skid along with little damage. Then comes smooth asphalt and finally cement. Stay away from turf or dirt if you can because you don't know when the prop, step, gear or flaps will dig in and really do damage," explained Geiger who was rapidly becoming an expert on such things.

Wbe he returned to Kerrville to pick up the ship after it had been repaired, he had his wife-to-be, Bonnie, along. They took off without incident and continued to Bonnie's mother's home in Newton, KS. When he moved the selector switch to "gear down," the same "pow!" happened again. The gear wouldn't come down with the emergency system. He flew to Wichita International where the tower confirmed that the gear was hanging free.

This time Geiger picked a long runway and built up plenty of speed, 120-130 knots. When he was sure he had the runway "made," he killed the engine and then "nudged" the prop with the starter until it was parallel to the ground. At the last possible moment, he retracted the flaps so that they wouldn't be damaged and used aileron on touchdown to counteract the off-center drag of the fixed step. This time, damage was confined to two pieces of aluminum "belly pan."

A phone call to the factory the next day came up with the request to get the gear down-and-welded and bring the ship back again. This time the factory checked the system more carefully, since another ferry pilot had just encountered the same difficulty. It was finally realized that the original gear extension system developed high forces with its rapid extension that forced the gear right on past the over-center position. The slower system caused the gear to "hang up," and after a few cycles, the old system would blow out.

The factory kept the twice-bent bird, replaced it with a new one and refunded Geiger's expenses for repair and travel. Perhaps that's why he sells an average of 45 new Mooneys each year.

She Runs the Business but Doesn't Fly

Aviation Services International (ASI) of Lock Haven, PA, ferries primarily Pipers, because of its location, and does considerable European delivery work (Fig. 3-6). Company President is Eileen L. Wolf who took over management of the company when her husband, Joe, was killed in the crash of a Piper Tomahawk whose magnetos failed over a swamp in North Carolina. Prior to that, Joe Wolf had made 35 to 40 overseas deliveries each year and worked the management end when at home. Mrs. Wolf, who is not a pilot, learned the business while keeping things running when her husband was "on a trip."

ASI requires that all their pilots have an instrument and multi-engine rating. Mrs. Wolf details: "Anything additional is so much the better. We prefer at least 2000 hours flight experience. This is subject to change depending on the ability of applicant. Naturally, previous ocean crossing experience, excluding X-airline and military experience, is preferred. It takes time to check out a new ferry pilot. This is a special pilot and is not every pilot's 'cup of tea'. We send a new pilot in a twin in company with another aircraft the first two or three crossings. How the pilot handles this experience is evaluated, and if we feel he is competent and ready for the next step, we then begin routine for single-engine deliveries. Most of our deliveries are single-engine aircraft.

"Pilots are required to have a waiver of the single-engine trans-Atlantic inspection at Moncton, N.B. This consists of the pilot making three consecutive stops with a single-engine aircraft in Moncton for inspection of the aircraft for required radio equipment, survival gear, etc., and questioning of the pilot by the inspector to determine competency. Once waiver is issued, pilot is cleared to leave on a ferry flight from Goose Bay, Gander and St. John's. We always try to send a number of single-engine aircraft out in company. Some problems can be solved by the buddy system, and it also helps morale.

"Our pilots deliver for us on a contract basis—a set fee for each model aircraft to each point of delivery. This amount is determined at the beginning of each new model year.

"We prefer to use the same crew continually. To this point we have been fortunate and have had present pilots one to three

Fig. 3-6. Piper retractable Arrow in the Monument Valley area. Many Pipers like this one are ferried overseas by professional delivery services.

years. You may be interested to know we have had a female making international deliveries for a year. Ms. O'Donnell could tell you of some 'surprised looks' when she steps out of the machines. There have been no negative reactions to her being on the crew.

"I'm sure any of our crew could tell you any number of unusual experiences. The best place to hear 'all about it' is Gander, Newfoundland, while a group of ferry pilots are waiting weather to make a crossing."

ASI's mailing envelopes have the notation, "Happiness is delivering aircraft."

The list of commercial ferry services changes. Many of the smaller companies are either one-man or on-woman organizations, while others have two or three pilots available when needed. Some of the larger ferry services are listed below and can be queried by letter for their specific qualifications:

Globe Aero Ltd.
Contact: Phil Waldman
P. O. Box 5775
Lakeland, FL 33803

Aircraft Facilities International
Contact: David Andrews
P. O. Box 18513
Wichita, KS 67218

Floair Inc.

Contact: W. K. Patterson
Hangar 17
Wichita Mid-Continent Airport
Wichita, KS 67209

Southern Cross Aviation

Contact: Peter Dickens
Municipal Airport
Santa Paula, CA 93060

Flynn's Flying Service

Contact: Tom Flynn
Newton City-County Airport
Newton, KS 67114

Aviation Services Int'l Inc.

17 Hickory Drive - P. O. Box 587
Lock Haven, PA 17745

Three Unusual Professionals

Most ferry flying is a lonesome, demanding and sometimes adventuresome business. Predictably, it attracts an unusual breed of pilots. Let's meet some of these outstanding pilots:

Louise Sacchi. Owner President Chief Pilot of Sacchi|Air Ferry Enterprises (SAFE) of Philadelphia, Louise Sacchi, made 200 transoceanic ferry flights in the first 10 years of her company (Fig. 3-7). Miss Sacchi, who enjoys fancy earrings and frilly blouses, wants to be called "Miss." "I don't like Ms. because that way they're neither fish, flesh nor good red herring."

Miss Sacchi, who claims to be "between 40 and 75 years of age," learned to fly in Long Island in 1939. Two years later, she completed a Merchant Marine course in celestial navigation and subsequently taught navigation to British Royal Air Force students.

The bulk of Miss Sacchi's deliveries have been to England, but she has flown new ships to Vienna, Finland and South Africa and has made 13 trips across the Pacific to the Philippines. She prefers the Atlantic because, "unless there is a typhoon or something, the Pacific is really pacific."

"In the Atlantic, there nearly always is some kind of weather; you rarely fly in sunshine. But those are the trips that make you tired. You sit there and all you have to do is switch tanks every once in a while. In the summer over the Atlantic, I can make a schedule for a customer. In the winter, I can't make that kind of a schedule. I arrive when I get there."

Crossing the Atlantic is a one-day job, while the Pacific takes at least four days.

Miss Sacchi also scotched two rumors that have begun to follow her:

"No, I don't take books along to read while flying alone—you're much too busy—and no, I don't take an alarm clock to wake me in case I fall asleep."

Part of Miss Sacchi's flight gear is a black sleep mask to take care of the various time zones. "When I get to where I'm going, I'm tired and ready to eat and get some sleep."

Her advice to would-be ferry pilots for transoceanic flights is, "Don't, there are too many of us now," adding, "a lot of people think they want to be ferry pilots until they get a look at the ocean."

See *Ocean Ocean Flying* by Louise Sacchi, published by McGraw-Hill, 1979, for more data.

Fig. 3-7. Louise Sacchi had 14,000 hours when she made her 86th Atlantic crossing (Courtesy Beech Aircraft).

Hilyer Dubois. North Atlantic crossings with single-engine aircraft are relatively routine. However, one that was out of the ordinary was the delivery of a two-place Piper Tomahawk to the Apache Flying Service in Manchester, England. Ferry pilot Hilyer Dubois of Ft. Lauderdale, FL, who just happened to bc 71 years-of-age at the time, broke some sort of a record with the tiny trainer (span 34 feet and length 23 feet). He operates his own delivery service and made the delivery for Red Aircraft in Ft. Lauderdale.

Dubois' normal pattern would have been to stop for refueling at Greenland and Iceland, "but when I went out to the airport to see if the weather was suitable, the conditions were so beautiful I decided to go straight to Shannon."

He said crossing the Atlantic in a small plane is not unusual. Dubois himself has made the trip "hundreds" of times since he started his delivery service 10 or 12 years ago.

"I never even gave it (setting a record) a thought until afterwards. I wasn't interested in records, I just wanted to deliver an airplane.

"I realized it when I got over there. It suddenly dawned on me—this is the smallest plane that has ever done it."

The company he was supposed to deliver the plane to recognized the record, too, and Dubois was greeted with television cameras and write-ups in the London press when he took the Tomahawk to Manchester.

Dubois said normally the limited fuel capacity of such a small plane would prevent it from crossing the Atlantic directly in one hop. Previously, the smallest plane that had made the trip was a four-placer.

He credits a 60-mph tail wind and near-perfect weather conditions over the stormy north Atlantic with enabling him to more than double the plane's normal range.

The "fantastic" tail wind was what made it possible; otherwise there's "no way" the 100 mph Tomahawk could carry enough gas to do it.

"I would have had an awful long swim without it (the wind)," Dubois said.

Based on the 24-hour weather forecast and his own verification of the speed and direction of the winds, Dubois began his trip without trepidation.

Although he makes radio contact every hour, "you don't know what kind of help you can get out there. If you're not pretty sure you can make it, you don't go." He also said airplanes are extremely reliable today and "it's rare to have an engine just quit cold."

The biggest problem he had to worry about was ice, common in the North Atlantic where temperatures hover close to freezing even near the surface.

Since the Tomahawk is relatively slow and has no de-icing equipment, "you have to be especially careful," even at 11,000 feet, which is low compared to the major airline's altitude of 30,000. Dubois said the temperature outside the heated cabin was minus 10.

But he didn't run into any weather systems until he got to within 175 miles of Shannon, where he knew he could drop down to warmer weather underneath if icing occurred.

"You could sit and wait for that beautiful pattern of wind and no particular weather system for months," Dubois said. "I was just plain lucky."

But sleep wasn't a problem. "You have a big incentive to stay awake. You're afraid to go to sleep," he said.

Total flight time from Gander to Shannon was 12 hours, 5 minutes.

Dubois said flying was something "I always wanted to do."

Fig. 3-8. Professor Art Scholl of Riverside, CA, checks an area with his Chipmunk. Note the smoke cartridge container under the wing.

After he got his license in 1940, he bought a plane for use in his business manufacturing and distributing specialty records to roller and skating rinks.

"I've flown almost everything built in the United States," he recalls, except for jet-propelled aircraft for which he is not qualified.

"As long as my health is good, I intend to continue flying," he said.

Art Scholl. "Ferrying is the most important single part of my air show career, whether it's for the shows themselves, films or TV commercials," explained veteran aerobatic pilot, Art Scholl (Fig 3-8). "I keep two Chipmunks ready to go—one on the east coast and one on the west—to use for the 35 weekends and 138 airshows I fly each year. Both airplanes come back to my home base each winter for overhaul, but they are parked between assignments during the summer."

Scholl does most of the ferrying himself because his two highly-specialized air show airplanes are irreplaceable. Each has been extensively modified to take repeated violent aerobatic demonstrations. At one time there was a theory that a pilot couldn't withstand more than seven positive and four negative G's. A NASA instrument in Scholl's Chipmunk recorded 12 positive and eight and one-half negative G's.

As a typical example of the problems in ferry flying, even by a veteran ATP pilot with all the ratings, 11,000 plus hours and a singular capability for low-altitude flying, try this for size:

Scholl flew a show in Wenatchee in Eastern Washington on a weekend and wanted urgently to get back to Southern California for a film commitment with the Chipmunk. At that time, United Airlines was on strike; so Art, his wife and their dog, "Aileron" departed for his home base at the Rialto Airport east of Los Angeles.

"There was that 'I've gotta-get-home' pressure," Art explained as he detailed the problems of ferry flying. "The first hop to Redmond, OR, was no problem. After a weather briefing that forecast zero-zero in 20 minutes, we headed south. For once, the weatherman was right; we turned back and circled up through a broken overcast to 8000 feet on top. Then, we headed on course toward Klamath Falls, OR, and had to climb to 18,000 feet to stay on top of the clouds. We couldn't maintain that altitude, so we called Seattle Center and circled down through a break in the clouds over a big lake just north of Klamath Falls. We tried to get under the weather, and I'm pretty good at low-level flying, but there was no way. The weather closed in—down to 50 feet—and we landed at a little duster strip north of our destination. We were so cold that we could hardly walk, but the people at the little store nearby were very friendly and took us in—even Aileron—to get thawed out and eat. We met another group of pilots who had already been there for three days waiting weather and arranged for a motel for the night.

"I checked weather and was told that it would break up before noon, so we slept in. When we got up, it was 0-0 with breaks and tops again reported at 18,000 feet. We took off, repeated our climb to 18,000 and couldn't make it south. I called the Center and told them that 'I did it again! The stuff is coming up all around us and if we're in any clouds, the engine will quit because of carburetor ice.'

"Sure enough, we went through a wisp of wet clouds and the Chipmunk quit. It was no sweat with all that altitude and a hole below, but the Center was most perturbed. We let down, kept talking to the Center, and spotted the same duster strip we'd taken off from earlier that morning. When the Center went out of VHF range, we switched over to Klamath Falls Radio and told them what we planned to do. As the temperature warmed up, it was possible to restart the engine, and we orbited the same strip for almost two hours, hunting for a break underneath tbe weather. There was no way and we spent a second night north of Klamath Falls."

Scholl took off at daybreak the next morning and was able to beat the building weather to Redding, CA, for fuel and a non-stop trip back to Rialto. Elapsed time was two days for just over 1000 miles.

Because the two Chipmunks are important each season far above their insured value, Scholl is extremely particular about who flies them. At the latest count, only Denver Airline Captain Vic Harper flies the unique retractable-gear Super Chipmunk, while both he and Chuck Wentworth, Aerobatic Instructor in Scholl's school in Rialto, trade off in the standard airplane.

"No matter how we prepare, weather is always a problem," explained Scholl. "Some FAA FSS personnel are 'weather readers only'; others are great. After flying all the way across the southeast to get to an isolated drag strip and fly an air show the next day, I let myself get crowded into a weather problem. I was flying the 'Super' and within 15 miles of my destination when the weather went down...and down...and down. I was flying along a freeway—that's what I usually do with these two new ships because I can't afford to have a crack-up—and had finally decided to land on the freeway. There was room and I was clear ahead. It was about dark and all of a sudden there were red lights ahead. It was a traffic signal, and I was already down below the wires. I firewalled it, went up and on instruments at once. I called the Center as soon as I was high enough. I explained that I had less than 45 minutes of fuel and asked for vectors to any airport that had some sort of a ceiling.

"I wouldn't have flown into these conditions if I hadn't had lousy weather briefing all day and an air show scheduled the following day.

"In any event, the Center picked a nearby airport that promptly went to zero-zero. I told the Center to vector me in anyhow for a precision radar approach with no go-around. I've forgotten the name of the airport—somewhere in Mississippi or Alabama. The FAA radar operator talked me down to minimums. It was dark as hell and raining like mad. Then he said, 'Missed approach,' and I told him 'Negative, keep talking.' I had the hatch open; it was wet and I was sweating.

"We came down to 100 feet and radar called 'Over the numbers—on the center line—now 1,000 feet down the runway—50 feet.' I kept going and didn't see a thing. Sure enough there was a little squeak of the tires and we were on! I never did see a runway light and it took a considerable time for a 'follow-me truck' to find the plane on the runway. That radar operator was mighty fine.

"Weather was lousy the next day and the show was canceled. I went home on the airline."

On one ferry flight to an air show in Florida, Vic Harper took off from California. After three or four days, he and the "Super" were only at Baton Rouge, LA. Harper decided to leave the ship there rather than wait another week on weather. Scholl went on later, picked up the Chipmunk and flew it to Florida for the show.

Since these two unique aircraft cannot be replaced and they are booked profitably for a full season each year, either Scholl or one of his pilots plans to ferry them along roads with landing places where they can be landed VFR at any time.

Art Scholl, known in air show circles as "the flying professor," has a very unusual aviation background. For 18 years, he was a professor in Aeronautics at the San Bernardino Valley College in Southern California. He has an associate's, a bachelor's, a master's and a doctorate in Aeronautics. He has flown almost 200 types of aircraft ranging from 1908 replicas up through gliders, balloons, jets and jumbo-jets.

At an airshow, Scholl and his Pennzoil Super Chipmunk perform grueling maneuvers such as the outside loop to ground level, and the lomcevak, in which the airplane tumbles end over end (Fig. 3-9). The name "lomcevak" is Czeck for "headache" and is a maneuver the Communist bloc countries outlawed because of its extreme risk to pilot and plane. Scholl does the lomcevak with red, white and blue smoke streaming from the wings and tail of his Chipmunk.

Since he is a qualified mechanic holding A&P (airframe and powerplant) and IA (inspection authorization) licenses, Scholl rebuilt and modified the original Chipmunk so extensively that the manufacturer no longer recognizes it as one of its own. The wing has had 18 inches cut from each side, the rudder has been enlarged, and a powerful 260-hp engine has been installed.

As a recap, it is interesting to note that ferry flying, of all the problems in air show coordination and precision exhibition flying, is the single most important part of Art Scholl's entire flying business.

An Actor's Lesson in Pilotage

A number of things made the delivery of Cessna Skylane N333GK from Wichita to the West Coast very unusual.

- The delivery flight was an owner delivery, but was all logged as cross-country dual instruction.
- The new airplane was NORDO (no radio).
- One of the ferry pilots was well-known in his own profession;
- The instructor's remuneration, besides expenses, was a magazine article and photos carried in Plane & Pilot Magazine.

"I want a special radio combination in my new Skylane. It'll be installed here in Van Nuys (California) and a ferry service will bring out the ship as soon as it's finished," explained George Kennedy, the owner-to-be. He's a big (6'-4"), enthusiastic new private pilot who then had 75 hours in his logbook. Since then he has owned an Aztec and a Skymaster.

He also is an Oscar-winning actor.

"It really isn't too difficult to fly from Wichita to Los Angeles without nav/com equipment as long as you wait for good weather," I commented. "Wouldn't it be good practice for you to bring the ship out yourself? Get an instructor to 'ride shotgun' and really learn the basics of that old-fashioned form of navigation called pilotage. That's map reading."

The result of this conversation put the two of us on the ramp at Cessna's Delivery Center, running a preflight inspection on Skylane N333GK. The "Golf Kilo" represents the initials of the new owner, George Kennedy. The triple-three was selected because "Fortunate things seem to happen to me in threes, even though I'd guess I'm no more superstitious than the next guy." (Fig. 3-10).

N333GK was a colorful Halloween-orange, black and white that the owner had requested. The panel was complete with blind flight instruments and an EGT. The only missing items were the black boxes that go down the center of the console. So, without any nav/com equipment,we would fly west just as it was done 30 years ago.

Since our destination was Van Nuys Airport, the third busiest landing pad in the nation, we did carry a small "no-name" VHF transceiver. However, the only time it worked was for close-in tower communication with Van Nuys at the conclusion of the trip. With this single exception, it was a remarkably quiet and restful delivery flight.

We stacked up the charts and climbed aboard. Kennedy explained that two of the main factors that led him to the Skylane were the adequate cabin size for his height and the ease of entry through either door. As it was, the two of us did a good job of filling the front of the airplane, but it wasn't cramped.

With the proud new owner in the left seat, we went through a complete by-the-book checklist before bounding into the blue skies over Kansas (Fig. 3-11). The recording tachometer showed 4.12 hours.

If you've ever picked up a new airplane at Cessna, you know that the departure pattern is one of the few legal buzz jobs in the United States. The Delivery Center is within the traffic pattern of McConnell Air Force Base, so all fly-away aircraft must take off and not climb higher than 300 feet until five miles east of the field. It's always an interesting couple of minutes with a brand-new ship, one with which you're probably not completely familiar.

Since a 300-foot traffic pattern was something different for the new owner, it was natural for him to crowd the top limits of the restriction. After all, such shenanigans are frowned upon at Van Nuys and other high density airports. However, we were soon away from the military traffic and climbed to a relaxing altitude over the plains of Kansas.

Fig. 3-9. Art Scholl performs his famous inverted ribbon pick-up with one of his chipmunks.

The Sectional charts came out and we debated between heading for Liberal or Dodge City. Dodge City won out for a number of reasons—no control tower, a weather station on the field and a short first hop to look under the cowling "just in case" something parted company. Obviously, an extended cross-country flight without nav/com equipment will teach any new pilot a number of techniques that he hasn't used before. He'll learn the old-fashioned use of reading air marking signs, the names of towns on water towers or grain silos, even dip to minimum legal altitude for the name of a small midwestern town on a railroad depot. When you're fairly sure in advance what the name will be, it's surprising what small print you can "guesstimate" from 500 feet with bifocals corrected to 20-20.

During the first leg of our westward junket, we cruised at 500 feet over the flat plains and pulled up another 500 as the small towns passed by. There were many readable signs to confirm map-reading in this particular area. This happy condition doesn't exist, of course, as you get into higher country.

Ours was a leisurely fly-for-fun delivery flight. You can ferry a Skylane to the West Coast in a single day if you want to work at it, but this was planned as a three-day trip to explore the off-airways areas that stretch between Kansas and California. It was fortunate that we had budgeted this much time because there was quite a chunk of interesting weather awaiting us along the line—and no radios!

We watched the water skiers on Cheney Reservoir just west of Wichita. We "drove" along Highway 54 and admired the bright-red barns, modern farm equipment and colorful growing crops.

We continued down the highway-and-railroad (IFR on this flight means "I follow the roads") past Pratt and Greensburg where the airports showed up on the charts. A right turn near Bucklin and we had Dodge City in sight.

As we unfolded from 333GK and the Mahon Flying Service gas truck approached, the lineboy asked Kennedy: "Haven't you been in here before?" Kennedy shook his head. "Then didn't you used to fly out of Tulsa when I was there?" Again Kennedy shook his head. "Well, I'm *sure* I've seen you somewhere before," he puzzled.

Kennedy grinned and replied quietly, "I'm an actor. Possibly you've seen me in the movies." Then we headed for the Weather Bureau.

Later the lineboy asked me about Kennedy while the new Skylane owner was in the cafe. Yes, he's been in Hollywood for

Fig. 3-10. Actor George Kennedy shoots landings with his brand-new Skylane during a no-radio delivery flight to California.

Fig. 3-11. George Kennedy runs a careful pre-flight check on his brand-new Cessna Skylane, the first of a series of aircraft that the actor has owned.

many years. Yes, he won the Academy Award in 1968 for Best Supporting Actor in *Cool Hand Luke.* Other leading roles were in *Gaily Gaily, Bandolero, The Boston Strangler* and *The Guns of the Magnificent Seven.*

As far back as he can remember, Kennedy has been fascinated by aviation. "G-8 and His Battle Aces" took the place of comic books when Kennedy was a youngster.

Kennedy's comments on why it took him so long to begin to fly should be of particular interest to today's manufacturers and every FBO in the country. Despite his size and recognized acting excellence, George Kennedy is a shy person. This isn't a publicity man's figment. You can't fly with another pilot for the better part of three days without figuring out fairly well what makes him tick.

"In the very short time I've been around pilots and doing my own flying, I've found them to be some of the nicest people in the world. It's just a little difficult to break through the imaginary barrier and get to know them as people," explained the actor.

It was playing the role of Joe Patroni in Arthur Hailey's *Airport* that actually gave him that final shove to go out and learn to fly. Kennedy starred in each of the popular "Airport" films.

During our jet flight east, Kennedy explained that he had hung-around airports every spring for at least five years but no one ever came up and asked me if I wanted to fly. I had the feeling that pilots were some sort of gods and I didn't particularly want to make a fool out of myself by asking how to join their group.

"Finally I wandered into Golden West at Van Nuys and a mature instructor named Brian Brooks took perhaps 15 minutes to go out on the flight line and explain to me what light plane flying was all about. He asked if I'd like an introductory ride, but I'd already made up my mind to give it a try. Three months later I had a private license and now Brooks is making me sweat on instruments."

Kennedy met "the real Joe Patroni," Roy S. Davis, who was then TWA's general maintenance foreman at O'Hare in Chicago. When Kennedy admitted that he was interested in learning to fly, Davis quipped, "Well, what's keeping you?" In real life, Davis is on the short side, graying, and chain-smokes cigars. Kennedy has been described as "burly," has a crop of un-gray, unruly hair and doesn't smoke.

But back to the flight line in Dodge City. Weather was reported touch-and-go ahead. The route we'd hoped to take over Denver, to Aspen and Rifle—I still wonder what kind of a Rocky Mountain town Rifle is really like—to Grand Junction was completely clobbered by weather. We elected to head straight west along the section lines to Pueblo, CO, and then take a look ahead. Tentatively, our destination was Gunnison, CO, but we didn't make it.

We climbed to 8500 feet for a breath of cooler air and picked a circuitous route around a series of energetic thunderstorms. Kennedy winced a few times when bright lightning flashed through the turbulent gray skies. This is not quite the kind of weather usually found in the smog-bowl of Southern California.

Later, Kennedy added..."After reading the proofs on Don Downie's report of our three-day, no-radio trip from Wichita to Van Nuys, there's just one point where I think he made the understatement of the year. He reported that I 'winced' when we were in the vicinity of our first heavy lightning strike. Wince, hell! I damn near jumped halfway out of the airplane. The whole trip was a wonderful experience and a liberal education to me, particularly at this stage of my flying in-experience."

Over Pueblo, we climbed to 10,500 feet and continued over Highway 50 toward 11,312-foot-high Monarch Pass. Actually, a good Rand-McNally road map would have been as useful as the Sectionals. We flew as far into the Wet Mountains—and they were very wet at that time—as Canon City before it became evident that there wasn't a ghost of a chance of topping the Rocky Mountains on this particular weather-filled day.

We executed that famed 180-degree turn and looked back at Pueblo. Forget it! A small, but furious, cloudburst was disgorging itself over both town and airport.

Out of the corner of my eye, I noticed my 75-hour pilot—by now he was approaching 80 hours—dry his sweaty palms on his trousers. Don't get the idea that Kennedy is lacking in intestinal fortitude; far from it. However, I would have worried about him as we bounced through the wet turbulence if he hadn't shown a few signs of apprehension.

"Well, where do you think we should go from here?" I asked quietly. When there isn't any radio static in the cabin, you can talk without shouting.

Kennedy took a careful look at the weather, glanced lingeringly at the downpour over Pueblo and then said, "I'd guess that we'd better head back to the East or head southeast. Maybe Trinidad will be open."

Trinidad seemed like a good selection—no control tower, a Flight Service Station on the field, and improving weather in that general direction. We made it with no problems and tied down N333 GK before an approaching rainstorm arrived. Fortunately, the Ramada Inn had space and sent a car the 18 miles to pick us up. The next morning we had wheels back to the airport and neither driver would accept a tip. Mighty nice people.

Trinidad is an unusual town. There are 59 bars within the city limits and on a Friday night during the height of the tourist season, most seemed to be nearly empty.

There was plenty of time to "hangar fly" and detail the problems of the previous day's flight. At Trinidad, for example, we watched from the motel swimming pool as the wild weather developed and figured how we'd fly around it. Then the charts came out to show what problems might develop for tomorrow's flight. However, when you're going cross-country without nav/com,you don't pick a destination with any real assurance of getting there on schedule.

The next morning, we checked with the FSS. Bad news! One local pilot had just reported that Raton Pass (7834 feet high) was closed, so we followed the tried, true and accepted procedure of checking in with the local FBO who eyeballed the weather and advised, "Well, you might go out and take a look at Raton Pass. This weather is mighty unstable and what was closed an hour ago may be open when you get there. However, if you can't get through there, you can always circle back east where the terrain is lower and follow the railroad over to Branson."

After fighting a losing battle with the hungry local mosquitoes, we took off in a drizzle for a look at Raton Pass. You could see a tiny sliver of daylight filtering through the trees, but there was no assurance that this wasn't a "sucker hole" with another cloud bank beyond. We made another 180, this time a tight turn back toward higher ceilings.

"When you get into a canyon like this, always fly close to one side so you'll have room enough to turn around," I'd suggested. "Then, if you do decide to turn back, you'll have space. Be careful to make this a smoothly coordinated turn. We don't want to skid up into the bases of those clouds."

The new pilot did as suggested, wiped his palms and we drove down the railroad tracks in exactly the opposite direction from where we really wanted to go. I pointed out to Kennedy two simple navigational aids he had with him that matched his Sectional chart: his watch and his pen. A standard ballpoint pen scales out exactly 40 miles on a Sectional. That's 16 minutes per pen-length at 2½ miles-per-minute. Sure, we had pilot-type plotters aboard, but the ballpoint-pen approach worked out with all the accuracy we needed and gave us more time to look outside the cockpit to compare landmarks on the ground with those on the charts.

There was a safe gap below the cloud deck to turn south over Branson, on the Colorado-New Mexico border. We headed toward Las Vegas, NM, and were able to see the tracks of the old Sante Fe Trail near Wagon Mound. This hazy imprint of the old wild west was of particular interest to Kennedy because of the number of Western films in which he has worked (Fig. 3-12). After looking at this remnant of yester-year transportation, all that the new pilot said was, "Wow!"

We turned the corner of the Sangre de Cristo Mountains, circled Sante Fe and landed. Flight time was 1 hour and 45 minutes—probably an all-time record for the most time in the air between Trinidad and Santa Fe with a plane that cruises as fast as the Skylane. The direct airline distance is only 146 miles, but we flew much farther.

Since the direct route between Santa Fe and Gallup is over lots and lots of very little, we filed a flight plan with the combined Tower/FSS at Santa Fe. Again we circumvented some very powerful, very wet-bottomed thunderstorms—tops estimated at over 30,000 feet—and cruised at 10,500 feet just south of the Chaco Mesa.

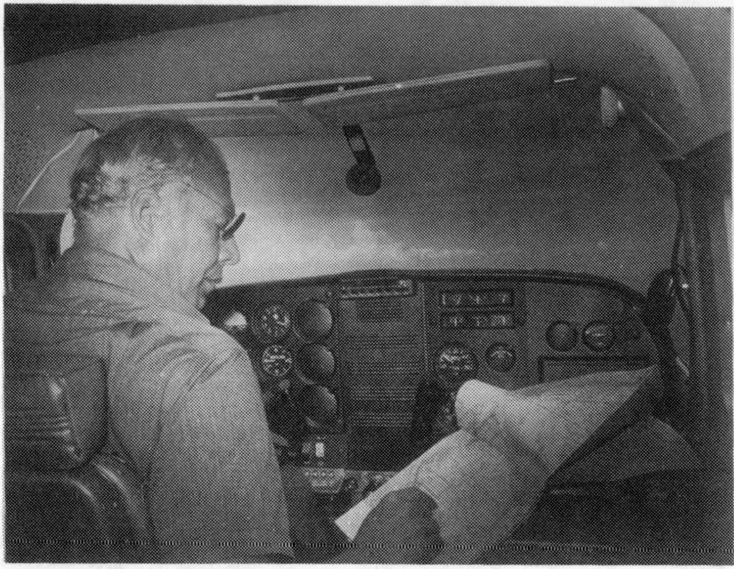

Fig. 3-12. Note the bare nav/com panel as George Kennedy studies a Sectional chart at 10,300 feet somewhere over the Western Desert where he had made several "horse operas". Radios were added after the Skylane was delivered to California.

Approaching Gallup, Kennedy applied carburetor heat, per factory instructions, and reduced power for a let down. Adding carburetor heat made the effective fuel-air ratio so rich that the engine ran quite rough. This was corrected by leaning the mixture control part way. When the new owner applied power again, he pushed off the carb heat—and things became very quiet. The mixture setting was too lean to sustain power, but a quick enrichening job had the big 230 Continental horses again at full gallop—approaching Gallup. Sometimes that cruising at 10,500 feet without oxygen sneaks up on you.

We topped the tanks and headed between the thunderstorms toward the technicolor wonders of the High Desert, crossed the Canyon de Chelly National Monument near Chinle and followed a wet, dirt road to Kayenta and on over the towering sandstone spires near Monument Valley.

Our leisurely flight planning had included the possibility of a stop at the Monument Valley Trading Post. We circled the area and could see a long series of puddles on the usually-dry dirt runway. Why get this spick-and-span new bird all muddy if we didn't have to land?

Fig. 3-13. Page, AZ, Airport Manager Royce Knight and George Kennedy discuss high-desert flying techniques.

Kennedy marveled at the ever-changing panorama of almost unreal color as the Skylane gobbled up the miles westward toward Page, AZ. Once we had landed, it was old-home-week for Kennedy. He'd spent two months here on location during the shooting of *Bandolero* with James Stewart, Dean Martin and Raquel Welch. Kennedy was forced to do considerable horseback riding to fit the role. "The only things that horses like are other horses," he commented succinctly. "There's a quotation that applies to almost every actor who works in a Western," he grinned. "From a house to horse to chiropractor."

Airport Manager Royce Knight had supplied a 172 to fly Kennedy to Flagstaff at dawn during the filming so he could have a chiropractor put his back into shape after the big man had come out second-best with a horse (Fig. 3-13).

You really haven't lived until you've spent a Saturday night in Page, AZ. The one small theater has a single showing of a single feature and all that remains is TV or boozing. Since hangovers and turbulent air make a poor combination, we watched TV and called it a night at an early hour. Kennedy freely admitted that he was quite tired after five hours of pilotage in some very rough weather and at altitudes up to 10,500 feet. So was I.

We checked weather the next morning with a no-toll line—ask the operator for Enterprise 675—with Bryce Canyon FAA/FSS. It was more of the same: scattered thunderstorms and increasing turbulence as the day wore on.

After nearly two days of working at this ancient form of navigation, Kennedy was due for a little relaxing flight. I suggested that he shoot a few landings solo in the calm early morning air over Page. While he was flying, I wandered out beside the runway and shot a few pictures of the shiny new Skylane.

Royce Knight was just buttoning-up the cowling of one of his Skylanes and I asked him if he'd have time to make a short photo flight. He had the time. Since Kennedy is an avid amateur photographer, I asked if he'd mind shooting a set of air-to-air photos with Knight while I "parked alongside" with 333GK. Kennedy agreed that even though he now had more than 80 hours, he wasn't yet ready to fly formation; so I chased Knight for a few minutes while Kennedy preserved his new bird in film shot through the open window of the older Skylane (Fig. 3-14).

We topped off on fuel and headed for the Grand Canyon Airport. Again, navigation was quite simple. We merely headed toward that great big ditch that spans Northern Arizona and flew west.

"Tell me if you think it would be safe for us to fly down into the canyon?" asked Kennedy. "Ever since I saw the first 'Cinerama', I've been intrigued with that idea, but if it sounds foolhardy, just say so."

When you've been flying for years, you develop into a professional coward, but the weather was calm and cloudy so turbulence would be no problem.

"It's fine by me," I replied, "with just one comment. Let's wait until we come to the intersection (when you're talking about rivers, the technical word is word is 'confluence') of the Little Colorado and the main canyon. There's a sand bar down there, if the water isn't too high, where you'd be able to land this ship just in case it quit. This is the same spot where a Cherokee Six was landed safely a year ago. However, if we do go in there, the only

Fig. 3-14. George Kennedy "shoots" his new airplane over the desert near Page, AZ, from a plane flown by Royce Knight.

way you'll get this airplane out is to disassemble it and bring it out by helicopter. That's how the Cherokee Six came out."

Under the conditions outlined above, I felt that this brief tour a couple of thousand feet down into the canyon was perfectly safe. However, if you're over this area when it's windy and there's turbulence, don't (even for a minute) consider getting out over that chasm.

The big Continental went into anticipated "automatic rough" just as soon as we dipped down over the canyon walls, but the view was superb. The water level was low, so our ace-in-the-hole sandbar was even longer than normal.

We leisurely spiraled up past the stony look-out at Desert View. Much of the rim of the canyon was studded with unseasonal patches of fog that would accompany the thunderstorms slated to develop later in the day (Fig. 3-15).

From the Grand Canyon, we headed out over the lonesome Coconino Plateau and crossed the Grand Canyon twice on a direct line to Las Vegas, NV. Lake Mead came into view on schedule and we tried our low-power transceiver on McCarran Tower. Nobody answered, not even when we were within five miles of the airport.

"So what do we do now?" I asked Kennedy.

He looked at the chart and replied, "Can't we go into North Las Vegas without radio? There's a Unicom on the field, but nothing in the regulations says we have to use it." Later a tower was added here and a TCA blossomed over the main terminal.

We landed at North Las Vegas and gobbled down a cold sandwich. One of the local pilots asked casually, "Didn't I see you shooting up a bunch of Indians on the screen last night?"

Kennedy returned the question: "Do you mean *The Guns of the Magnificent Seven* has been playing here? I didn't know it was released yet."

We walked out into the searing Nevada sun for a final tussle with the thunderheads. The weather was just as forecast. The cumulus build-ups were scattered, but the granddaddies went well above 35,000 feet. A king-sized cloudburst was in progress east of our course over the Providence Mountains.

The weather ahead looked anything but promising as we came up over the Daggett Airport with its excellent FAA/FSS. This was the only time on the entire trip that I wished we had good communications equipment aboard. It would have been very convenient to have known what the weather was like at that time in Cajon Pass, at Ontario International Airport, at Van Nuys itself, and to find out if there were any PIREPS (in-flight pilot reports) on the weather in both Cajon Pass and Mint Canyon to the northwest.

Fig. 3-15. Arizona's Grand Canyon near Desert View as patchy clouds obscure part of the gorge. Much of the rim of the canyon was obscured with moisture that would accompany thunderstorms later in the day.

However, without usable radios, the only thing we could do was bore into the haze with the programmed thought of making another 180 at any time the weather really became a problem.

There are four passes leading into the Los Angeles Basin from the desert: Banning, Cajon, Newhall and Mint Canyon. The normal route from Las Vegas to Van Nuys comes over Palmdale and skirts Newhall Pass, but what had really been normal on this entire trip?

As Daggett FSS silently dropped behind us, I asked Kennedy how he planned to handle the weather ahead. He indicated Mint Canyon and Newhall Pass as probably the best route, and there's a good chance that he was correct. However, I suggested that we go in and take a good look at Cajon Pass, since it was directly in front of us anyhow. If we could see enough through that one, we could turn the corner of the San Gabriel hills while letting down into the smog.

A second, and purely personal, reason for exploring this approach was that I've lived in the San Gabriel Valley for years and know the location of the airports, the freeways and the TV towers. Therefore, eyeball navigation would be easier for me from this hometown direction.

We turned the corner of the Cajon Pass and headed into the heavy haze, studded with tiring cumulus, toward Van Nuys. Soon we were in Kennedy's hometown traffic pattern, so he took us over the Sepulveda Reservoir for a try with our on-again/off-again communications unit.

Surprise, it worked! Van Nuys tower came back with a cheerful, "Roger Cessna triple-three-Golf-Kilo, you are cleared straight in on one six right. Report the Hospital (two miles out on final approach)."

N333GK promptly went into the shop for two nav/com sets, transponder, DME and ADF and a Carb air-temp gauge. Kennedy went home with 12½ hours of old-fashioned, grass roots pilotage experience in his log book. If and when a nav/com set blows a transistor, he'll be able to handle the problem without sweaty palms.

After all, he's been there before.

Chapter 4

Shiny and Ready-To-Go

No matter how often you've done it, there's always something special about the delivery of a brand-new plane. As a pilot, you associate the names of places like Wichita and Strother, Vero Beach and Lock Haven, Albany, Oklahoma City, Savannah and Kerrville with the beginning of some mighty interesting flying. In earlier days, some of these new planes came from Los Angeles, Long Beach, San Diego, Cleveland and Middletown.

There's little rhyme or reason for where aircraft factories locate. Wichita, KS, is known as the "home of general aviation" through the historic hometown efforts of Clyde Cessna, Walter Beech and other pioneers. The Piper factory in Lock Haven, PA was located on the banks of the Susquehanna River only after the original factory in Bradford went up in flames in 1937. For a ferry pilot, Lock Haven is still one of the most difficult factories to reach (Fig. 4-1).

Piper facilities in Vero Beach and Lakeland, FL, were results of the continuing growth of general-aviation aircraft. Cessna's satellite factory at Strother, KS, indicates a similar growth in production.

After a number of moves, Mooney has settled in Kerrville, TX. Rockwell Aero Commander has production facilities in Oklahoma City, OK and Albany, GA. At the moment, Grummans come from Atlanta, GA. In the north-central section, Bellanca bases at Alexandria, MN, with the popular Citabria coming out of the Os-

Fig. 4-1. One of the newer breed of "T-tail trainers," a Beech Skipper, bores holes in the sky near Liberal, KS, where the trainer is manufactured. Liberal is not the easiest place to reach on scheduled airlines and several dealers prefer to fly one of their own four-placers to the factory and pick up three new aircraft at a time (Courtesy Beech Aircraft).

ceola, WI, division. On the West Coast, Piper's Aero Star division is in Santa Maria, CA, and George Varga's Kachina is located at Chandler, AZ. The two-a-month Pitts aerobatic jewels spawn at Afton, WY.

From a ferry pilot's viewpoint, how the heck do you get to most of these locations? Wichita is no problem on the airline; Strother is something else. Vero Beach, FL, is an on-and-off airline stop with most ferry pilots coming by airline to Melbourne, FL, and working their way from there. Lock Haven, PA has been an isolated destination ever since it began; you can get to Williamsport on an occasional airliner, but it isn't easy from there; it takes a cab, rent-a-car, or a friend at the factory.

One of the most difficult factories to reach has to be Citabria's spawning grounds in Osceola, WI (Fig. 4-2). It takes a large-scale map to find the place. There's no airline, no bus—nothing. Recently, ferry pilot Bob Weston picked up a 115-hp Citabria for delivery to the West Coast. As he tells it, he rolled off the night coach just before dawn and took a local bus to the Trailways depot. Then another bus took him to Hudson and the end of the line. It was a question of using his thumb on the highway on a very chilly morning or paying $20 to the Hudson Cab Company. He took a cab to the factory door in Osceola where Chief Pilot Dick Johnson provided a car to get to town for a late breakfast before taking off.

With the volume of deliveries from Osceola, plus the advantage of a "working press pass" that is supposed to admit you to "the back door of hell," I've been fortunate with Citabria. At one time, a nearby FBO would meet your airline flight at Minneapolis and ferry one or more to Osceola in a C-172 for $25—then inflation set in. However, the ferry flight was well worth the effort, considering the per diem costs in good weather. There have been Citabria deliveries where factory test pilot Dick Johnson has gone to the special effort to bring the new ship down to the St. Paul airport, Holman Field, and leave it in a heated hangar when the weather was freezing. That makes a great combination—most appreciated, but not easily arranged by the factory.

When there is more than one ferry pilot, the economics of using a rental car are attractive. From Melbourne, FL, for example, you can usually arrange to pick up a car and leave it at Vero Beach. However, if you're going to Afton, WY—(where the Call Airs were originally built)—to get a new Pitts, you'll go in by bus from Salt Lake City unless you want to charter a trip. Usually, the dealer, FBO or eventual customer is fairly familiar with local connections; but if you're on your own, lots of luck.

On one Citabria pick-up, I went east with the legendary Bob Said, former editor of *Private Pilot,* and later *Plane & Pilot,* and "discoverer" of "Jonathan Livingston Seagull." As editor of the old Private Pilot Magazine, Said received an original manuscript from Richard Bach, enjoyed it, set it up to run and contacted the author for more material. "Bach told me that every aviation editor in the country had turned down that article and that if I hadn't run it, he was going to burn it," reminisced Said. "The first article I ran and the two more he wrote for me comprised the complete 'Seagull' book." (See Fig. 4-3) Said was taking the first of the turbo-charged 360-hp Comanches back to Lock Haven and offered to drop us at Osceola. The new Piper turbo installation had an original problem with hot starts, and I can remember unloading baggage and cameras at Osceola with Said in the cockpit and the engine running. It had taken us a half hour and a starting cart to get the "beastie" going after our last fuel stop in St. Paul. From Osceola, the editor continued to Lock Haven and we headed west with a 150 Citabria on a leisurely junket that went all the way to Banff, B.C., before returning to Southern California.

On two Piper deliveries from Florida, we were able to obtain a "working press" ticket on the Los Angeles Dodgers' Boeing 720

Fig. 4-2. Citabria Decathlon with ferry pilot Jack Switzer up front. On this particular delivery flight, Switzer was away from Southern California just three nights—one on the "red eye" to Minneapolis and two additional nights boxed by weather in Las Vegas, NM.

Fig. 4-3. Former aviation editor Bob Said with a copy of the original Jonathan Livingston Seagull article that he published in the old "Private Pilot Magazine."

from Captain Lewis Carlisle in exchange for material in print. It's a great way to reach Vero Beach from the West Coast—four hours non-stop and no cabs at the end of the line!

From a personal standpoint, I've always felt that once I've signed for the new airplane and have my bags and charts aboard, the hard work is completed and the fun is about to begin.

The First Time

You'll always remember your first trip to an aircraft factory. Mine was in Lock Haven, PA, during the era of the ubiquitous yellow J-3 Cub. It included a full factory tour before signing for the airplane and heading west. Whatever the factory, try to plan for a tour of the assembly line to really appreciate the complexity of today's aircraft assembly.

Since Cessna produces the largest volume of new airplanes, you'll probably visit Kansas before you get very far into ferry flying. So many ferry pilots pass through the swinging doors of Cessna's Pawnee Division Delivery Center in Wichita that there's a special "Arrival and Departure Information" booklet for the newcomer. I've been through this route with 16 Cessna deliveries plus several factory press demonstration flights (Fig. 4-4).

For first-time visitors, we'll reprint Cessna's booklet verbatim so that you won't have the usual ferry pilot's dilemma of reading the owner's manual or operational handbook on a brand-

Fig. 4-4. Aerial view of Cessna's Delivery Center shown in the foreground. Parked aircraft ready for delivery are at bottom. Aircraft that have been test-hopped or are awaiting test-hopping are in the middle of the grassy field. The main Cessna plant is at the top (courtesy Cessna Aircraft).

new airplane after you've already flown it for a day or two. Late-night motel reading of the aircraft manual usually elicits comments like, "Oh, that's what the little knob is for!" or, "Now I see why the fuel selector valve goes that way!"

Just as with the owners manual, factory fly-away instructions are best digested before starting the trip.

Arrival and Departure Information
Cessna Aircraft Company
Pawnee Division Delivery Center

Foreward

Welcome to the Cessna Pawnee Division Delivery Center. The Delivery Center Supervisor and Staff wish you a pleasant visit!

To make your visit by air as simple and safe as possible, this booklet has been prepared to provide you some easy-to-understand rules that must be followed during your arrival and departure. Whether this is your first visit to the Delivery Center, or you may have been here many times, we believe this booklet contains some facts you will find helpful, and we urge you to take a few minutes to study its contents.

Airspace

As you will note from the map (McConnell Air Force Base Traffic Pattern) the Pawnee Division Delivery Center is located on the perimeter of McConnell Air Force Base (see Fig. 4-5), and is

Fig. 4-5. Cessna's narrow delivery runway with rows of new aircraft is at the left. A portion of McConnell Air Force Base can be seen at the top. Most deliveries take off to the north with a sharp, low right turn out (courtesy Cessna Aircraft).

just three miles south of the Beech Aircraft Corporation Airport. This results in some rather congested airspace at times.

We have shared the airspace with McConnell and Beech for a number of years and have developed an outstanding flight-safety record by requiring strict adherence to our designated flight levels and traffic patterns, and through the continuing vigilance of our visitors.

While our McConnell Air Force Base neighbors fly a very regimented traffic pattern, their aircraft design and flight configuration during takeoff and approach to landing make their peripheral vision very limited, and an aircraft straying into their flight pattern could be in a very hazardous position. We em-

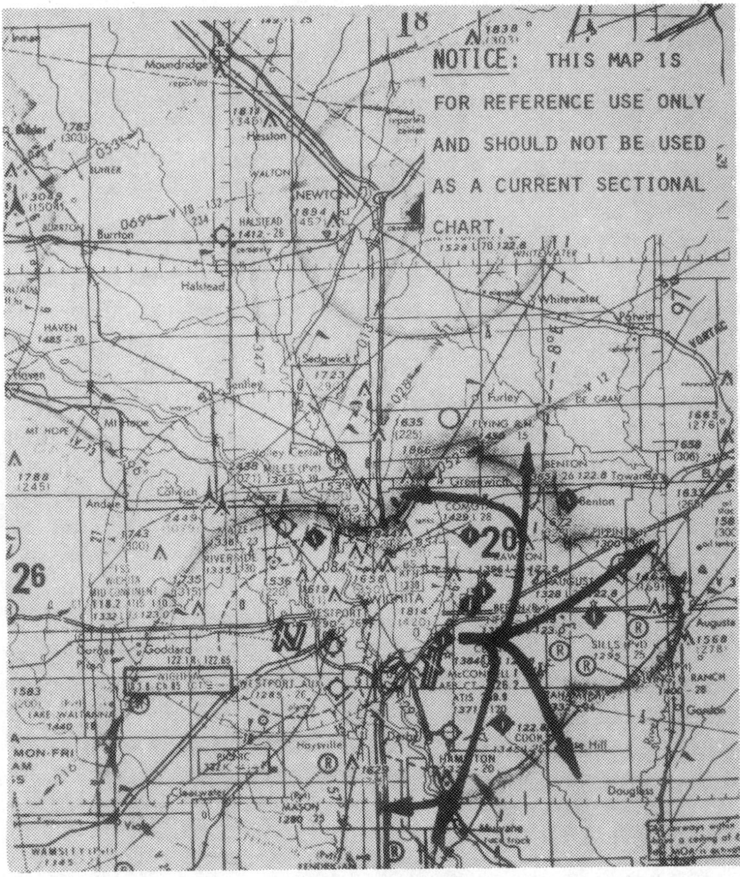

Fig. 4-6. A copy of the Wichita Sectional chart shows Cessna departure patterns. All traffic departs east of the factory.

phasize, however, this is no problem for you provided our designated traffic patterns are followed (Fig. 4-6).

Flight Levels

The Pawnee Division Delivery Center bas been assigned a flight level of 1700 feet MSL (300 feet AGL) within one and one-half miles of McConnell Air Force Base, and no more than 1900 feet MSL (500 feet AGL) within five miles of McConnell. This provides a vertical separation of 700 feet and 900 feet, respectively, from any military aircraft which are in their traffic pattern at the same time you are in the Cessna pattern.

Once beyond the five mile limit, additional altitude may be added; however, it should be added carefully in the north or south approach to McConnell, since their aircraft may be making a long straight-in approach or departure.

Traffic Patterns

The chart on McConnell Air Force Base traffic pattern illustrates the recommended Cessna traffic pattern along with the standard McConnell traffic pattern.(Fig. pattern (Fig. 4-7). Due to the proximity of McConnell Air Force Base and the Beech Airport, no straight-in approaches or departures are allowed at Cessna from either north or south, and *no turns to the west* are allowed as this would take your flight path directly across the McConnell Air Force Base traffic pattern. Just remember this simple rule: *All arrivals and departures at Cessna must be to and from the East.*

When departing to the south (Runway 17L), a left turn should be executed near the end of the runway, and when departing to the north (Runway 35R), a right turn out should be executed near the end of the runway.

Runway Designation

The Delivery Center runway is designated 17 Left and 35 Right. There is only one runway at the Delivery Center, but the left and right designation is required since there is another runway at Cessna located near the main plant, approximately one-half mile west of the Delivery Center Runway. This runway is used for flight test only and *is closed to the public* (Fig. 4-8).

Communications

The Pawnee Division Delivery Center transmits and receives on the Unicom frequency of 122.8. All aircraft arriving or depart-

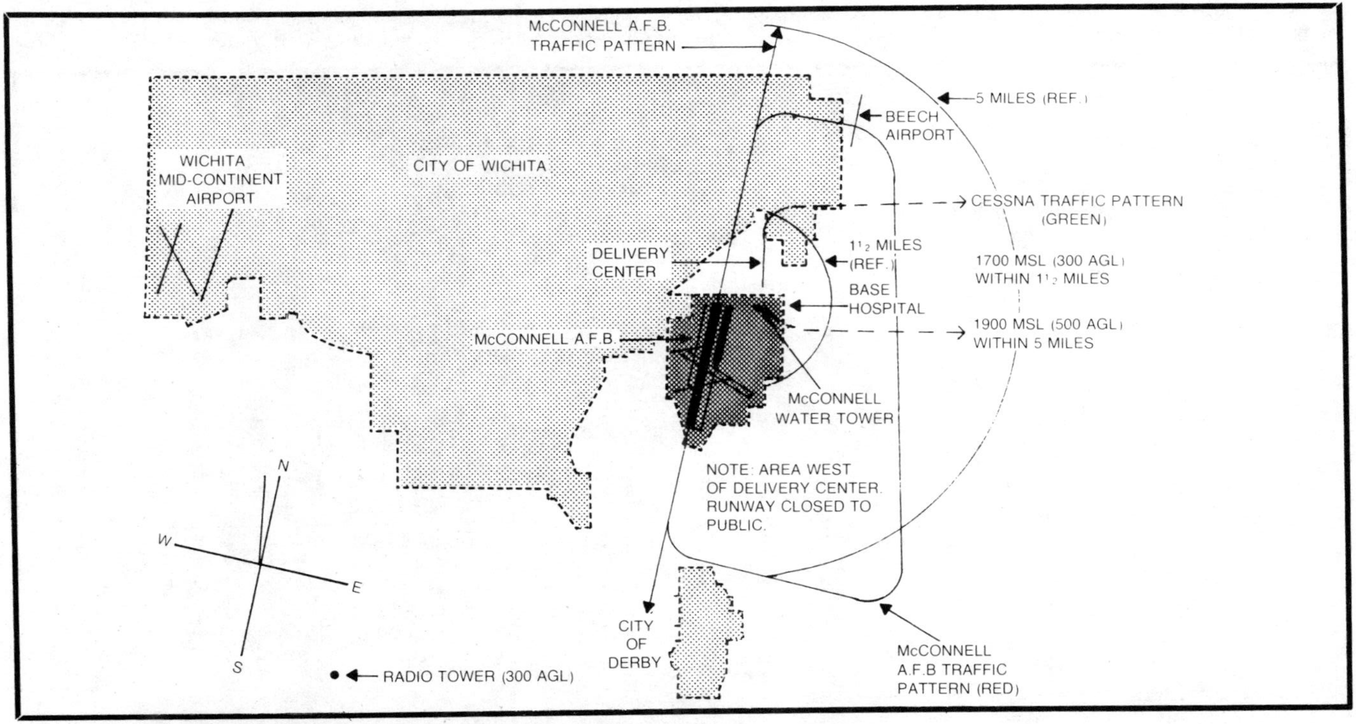

Fig. 4-7. Traffic patterns, in detail, for the Cessna Delivery Center.

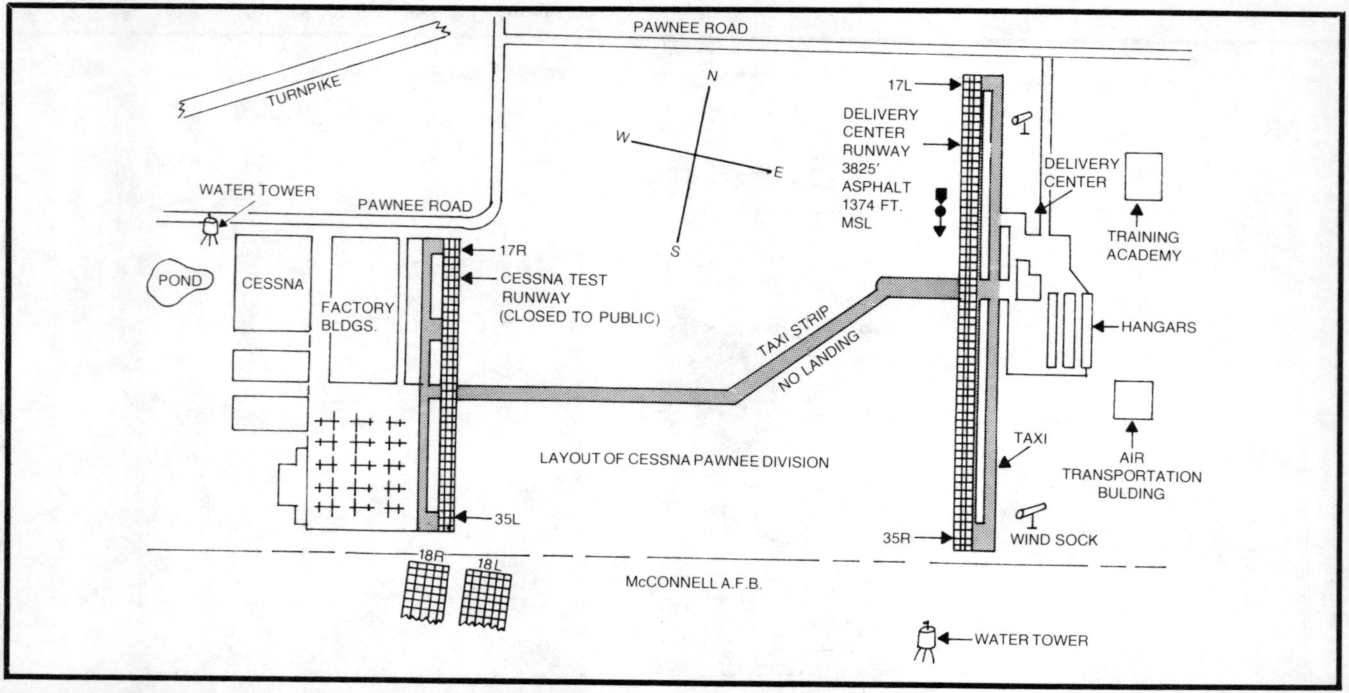

Fig. 4-8. Close-up diagram shows the main 17L-35R departure runway at Cessna. The shorter runway adjoining the factory is used only for test-hops and is not open to the public. No landings are permitted on the taxi strip that bisects the area between factory and delivery center.

ing are requested to call "Cessna Unicom" on this frequency for an airport advisory. Up-to-the-minute weather information is provided by telewriter from McConnell Air Force Base and will be relayed upon request.

A convenient 3x5 card containing most of the local radio frequencies is available upon request at the Delivery Center counter.

Flight over Adjacent Housing

To avoid unnecessary annoyance to our neighbors near the Delivery Center, such as the McConnell Air Force Base Hospital (which is an H-shaped building with a red cross on its roof near the east perimeter of the base), we request that all arriving and departing pilots select a route that will avoid flight directly over buildings whenever possible. Remember, your traffic altitude is only 300 feet above the ground.

Recommended Departure Routes

We receive many questions asking what the best route is away from the Delivery Center, and to assist in answering these questions, we have included a map of tbe Wichita area which has several recommended departure routes marked on it. This map is for reference purposes only and will not be up-dated.

Delivery Center Operating Hours

Normal operating hours for the Pawnee Division Delivery Center are: 8:00 a.m. to 4:30 p.m. weekdays, and 8:00 a.m. to 1:00 p.m. on Saturdays. We are closed Sundays and holidays unless otherwise announced. Our Unicom communications frequency is operative only during our operating hours.

We hope you enjoy your visit to the Cessna Pawnee Division Delivery Center. If you have any questions regarding aircraft pick-up or departure briefing, please ask one of our personnel for assistance.

None of the other factories that I've visited have the airspace constrictions of Cessna's main factory in Wichita. Most all others are located adjoining large municipal airports. Some have towers and others use a Unicom frequency for traffic advisories. You can get this information when you sign for the airplane.

Log books catalogue a variety of memories. These are a few of my personal experiences while ferrying new aircraft around the country in single engines and twins.

Piper's Retractable Arrow

When Piper first produced the retractable Arrow, we had the opportunity to bring the first one west. As with any first production airplane, things tend to be different. The Arrow has a goof-proof landing gear that extends by itself when either the manifold pressure gets too low or the airspeed drops.

N7308T presented a number of challenges that have since been ironed out. There were no gear-up locks on this airplane. Therefore, if you lost hydraulic pressure, the gear would fall out and there was nothing you could do about it. How, then, could you figure the plane's performance at altitude, on top or at night? With the gear down, you weren't going to go very far or very fast.

We were faced with this uncertainty at Tucumcari, NM, where a narrow layer of clouds hugged the Continental Divide. At least eight airplanes were headed west, and we held a session of the FCC (Flying Cowards Club) in the FSS. One National Guard

Fig. 4-9. In-flight cockpit study of one of the original Piper Arrows. Altitude is 8,500 feet with IAS 120 nm. Note trusty Sectional chart over the pilot's left knee.

Fig. 4-10. One of the early Piper Arrows in high-altitude flight near Flagstaff, AZ. The original model has some problems with gear retraction (courtesy Tom Roberts).

L-19 made a stab at it underneath and came back. Santa Fe, NM, not too far west, was reporting clear. The Flight on top would be no problem unless we had a hydraulic failure and the gear extended automatically. Under these conditions, you'd never be able to maintain altitude on top. We waited.

The original Arrow (Fig. 4-9) had been tested only under simulated high-altitude conditions. Under actual conditions at Las Vegas, NM, and Flagstaff, AZ, it didn't do one bit of good to put the landing gear in the "up" position after takeoff. There just wasn't enough high-density breeze coming across the sensing vane of the gear activator to let the gear come up. A mechanical override in the cockpit was required on every high-altitude takeoff we made in order to get enough airspeed for the gear to stay up by itself (Fig. 4-10).

Later pilots found that a Jeppesen book was just the right dimension to hold this override lever in the "up" position.

Eventually, the Arrow became a winner for Piper, but we offered to run a real-time independent flight evaluation for the company at Flagstaff after delivering the first ship. Vero Beach didn't go for the idea.

However, their advertising agency did come up with an immediate assignment for color photos of a new red Arrow in the Far West. Pilotphotographer Tom Roberts went along for a tour of the colorful sandstone cliffs at Monument Valley over Lake Mead, and later at 18,000 feet over Mt. Whitney. Actually, that model Arrow wouldn't go to that altitude, but we encountered one huge thermal while climbing up the face of the snow-capped mountain and circled. After the second go-around, Bob White, who was flying Tom Roberts in his Cessna 180, radioed across, "Have you got any oxygen?" The reply was, "Negative. Let's make one more 360 and get our pictures."

Roberts leaned out the window of the rugged C-180 as we made a leisurely circle over the largest peak in the contiguous states. His lenses and timing were sharp and the resulting photos were used on magazine covers and in national advertising for more than a year.

Then, at 18,000 feet, we dropped the nose to 130 mph indicated (that's about 177 mph true), and the doggoned gear fell out from under us.

Research along the Border

On one Vero Beach delivery we took a Cherokee to several of the airports where early dope smuggling was reported. Only with a previous arrangement with U.S. Customs Officials, that they give prior approval to the subsequent AOPA Pilot Magazine article, were we able to talk with narcotics agents all the way across the country. The conclusions: here is a small, dedicated group of officers that you'd really want to have on your side (Fig. 4-11).

Wings for a Racing Driver

On another delivery, this time an Aero Commander 200 destined for the Indianapolis 500 and driver A. J. Foyt, we picked up the almost-new ship from the Brittan Auto Pilot Shop near the Hawthorne, CA, airport. It had just that morning come out of an FAA flight test approving the new system, and Brittan factory engineers were lounging around the hangar to make sure that we got into the air safely. This is not really a good confidence maneuver, but Bob Weston and I took off in mid-afternoon for Oklahoma City where a factory pilot would take the ship on to the factory for final painting. It was another of those "rush-rush" things.

The 285-hp Aero Commander was fast; but without a turbocharger, we couldn't get above the afternoon cumulus build-ups across the desert. By the time we approached Flagstaff, AZ, we

Fig. 4-11. Beech AT-11 confiscated for marijuana smuggling at Nogales, AZ. It is strongly recommended that flights near U. S. borders not include unnecessary shopping.

were getting well beaten up and quite green under the beltline, though neither of us admitted it until we finally decided to call it a day and make Oklahoma City early the next morning. With both of us, cameras and bags and full fuel, we were approaching gross weight, but our ground speed eastbound with a 30-knot tail wind was spectacular. We had planned a breakfast stop at Tucumcari, NM, but east of Albuquerque, we noticed a slight discharge on the plane's ammeter needle. A placard on the panel indicated that this was undesirable and an inspection was in order. However, the situation didn't get any worse. Albuquerque was behind us and upwind; Tucumcari had only a small mechanical shop. Amarillo would have had adequate service facilities, but the ship was running well and it wasn't all that much farther to Oklahoma City where an Aero Commander factory pilot was waiting. We shut off the master switch and elected to continue with pilotage to Oklahoma City and then draw electrical power only for landing communications. Our total flight time from Flagstaff was a scorching 3:52, which isn't at all shabby for a single-engine plane at lower altitudes.

After landing, we explained the ammeter problem to the factory pilot and dropped the problem in his lap. Neither of us ever did find out whether the problem was in the electrical gauge or what. To continue the flight was a judgment call and each of these unusual situations calls for its own set of answers.

Excess Baggage: Check Your Weight and Balance

After visiting the AOPA offices in Washington, D.C. one summer, the only thing I could find to ferry west was an early two-place American Yankee out of Cleveland. The trainer delivery saved an airline ticket and offered some interesting flying. When I made my first takeoff in N5720L at near-sea-level from what was then the Yankee factory, I very nearly circled the field and returned the airplane. Take-off roll was long and rate of climb was marginal, even though I was alone in the plane.

The factory public-relations man had asked prior to takeoff if I'd mind, "taking a few brochures along to the new dealer in Ontario, CA." I agreed and promptly forgot it. Two days later, while looking for something in my bag in the baggage compartment, I found that the "few brochures" weighed 78 pounds. I knew that it was my responsibility for this weight-and-balance error. The brochures went on by truck and the little trainer was able to get across the mountains via El Paso, TX.

Murphy's Law

Even with a current checkout in a Bellanca Viking, I was able to make the same mistake twice on takeoff from Alexandria and continue to write about it. The Viking had three identical vernier knobs in the cockpit, all within a hand span. One was for throttle, one for prop and one for mixture. After flipping the gear-up switch, I came back on the mixture—*not* the throttle or the prop—until the engine began to spit. Of course, any good pilot knows that you're never supposed to move a power, fuel or accessory control before looking at it and identifying both function and proper movement.

Two Cylinders Are Almost Enough

On another trip, we met a fine helicopter mechanic at Cut Bank, MN, who listened to the 60-hp, two-cylinder engine on Citabria's economy $4995 Champ and was able to free-up one valve enough so that we could top the snow-capped Bitterroot Range. It seemed that one of his 'choppers had a Franklin powerplant with a similar valve arrangement. That particular airplane, N9135L (No. 3 off the production line) finally found its way to Anchorage, AK, still using the unique two-banger. Engine problems plagued this particular combination, and it is a rarity to see one still flying. If you hear a most novel engine barking overhead, there's a chance it may be this rare little Franklin.

Another Spur-of-the-Moment

One cold day just before Christmas, we were in Wichita doing a magazine article on Beech's new Sundowner. The only Beechcraft going west on delivery was already spoken for, and we checked to no avail with Cessna; then we scheduled an airliner home. Then came a phone call from Beech advising that the new owner's wife wasn't feeling well and they planned to return to California by jet. Sure, I'd take the new Musketeer Sierra, N9727Q, to the West Coast. This particular airplane had minor hail damage to the wing skins, so it had been sold at a discount rather than being re-skinned or having the various small "dings" filled and then being repainted. "Dings" or not, the retractable Musketeer was relatively fast, and I was at Beech West in Van Nuys 11.2 hours later (Fig. 4-12).

Citabrias Are Seldom Boring

No matter what happens on a new delivery, you can expect something a little different from each airplane. Take the little

Citabria with a loose tail wheel that would castor (spin) on a three-point landing—about half the time, it would throw one tail-wheel spring out in the middle of the runway. We tried safety wire, gun tape and nylon rope. The ship still ejected tail-wheel springs on rollout. Since only a few more landings were needed to get the ship to the dealer, where he could solve the problem at his leisure and in his own shop, we just brushed up on our wheel landings and lowered the tail gingerly at a minimum speed. It worked!

I delivered one Citabria with the left-wing root cavity taped over. On the "squawk sheet," I wrote: "One large bumblebee, probably dead, behind the tape and rag in the left-wing root." The huge, black bee had flown into the cockpit during an early fueling stop west of Osceola (Fig. 4-13). No amount of encouragement with folded maps, rags or other prods could get the bee away from the curved windshield. I don't think that any pilot enjoys killing anything that flies, except perhaps mosquitoes, gnats and that one noisy horsefly in the bedroom on a quiet night. However, our intruder was large enough to have inflicted both pain and possible temporary illness in case of a sting in flight. We borrowed an insecticide spray can, emptied part of it into the cockpit and waited. All that happened was that the bee appeared irritated. Finally, he backed his way into the wing root where we followed with more spray. The buzzing continued and perhaps increased in volume. The day was progressing, while we weren't. So we stuffed a rag into the recess of the wing root with the bee somewhere behind it. Then came a roll of gun tape to completely seal off the area.

The mechanic on delivery asked, "Do you think he's really dead?"

I shrugged and commented, "You'll probably find out."

Sometimes on a leisurely ferry flight, you'll detour by whim, rather than weather. It's interesting to actually land at places like Broken Bow, NE, Ft. Morgan at the entrance to Mobile Bay, AL, or Shelter Cove on the northern California coast. How about Plentywood, ND; Medicine Hat, WY; Hope, Salmon Arm and Chilliwack in Canada; Tuweep on the lip of the Grand Canyon, or nearby Dinosaur City? Ferry flying can be your key to "faraway places with strange-sounding names."

Twins Can Be Twice the Fun

A large percentage of domestic twin-engine deliveries are made either by clients who have just bought the airplanes or by

Fig. 4-12. Beechcraft Sierra headed west. Note that the nose gear has not retracted completely on this aircraft (courtesy Bob Beuttenbach).

Fig. 4-13. "There's a bumblebee somewhere in that left wing root," explains the author. "Now it's your airplane, unpeel the tape and see if he is still with us." Unhappy owner is Mike Dewey.

dealer-salesmen who have business at the factory. However, every so often there's a new twin waiting to go—both domestic and out of the country.

There wasn't a cloud in the sky from Wichita all the way to the West Coast, winds were negligible and I was in the driver's seat of a brand-new 1975 Beechcraft Baron B55 on my way to Los Angeles.

It was one of those days when it was almost sinful to be flying alone. Somebody else—up to five other "somebodies"—should have been sharing the beauties of a fine trip west in this speedster—sharing the new-plane smell and the glistening reflections of fresh paint, the unscuffed windshields, and a full-house panel of goodies that took me most of the afternoon to fully appreciate.

It had all begun with a phone call earlier in the day. I'd arrived in Wichita, delivered the Cessna Cardinal—that had been on display at AOPA's Plantation Party in Hollywood, Florida—back to Cessna's ramp and made an airline reservation back to Los Angeles. Then I checked the local airplane factories just in case anyone needed a delivery going west. A few minutes later, Dave Cotten, assistant director of public relations for Beech Aircraft Corp., called back and asked if I'd mind bringing out a new Baron for Mike Gordon at Beech West in Van Nuys, CA (Fig. 4-14).

Mind? I was delighted!

Twenty minutes later when Dave drove up, my bags and I were outside the Ramada Inn waiting. At Beech's delivery center, I signed for the ship, canceled my airline reservation with glee, checked the weather, and filed VFR to Albuquerque. AOPA Pilot Magazine originally carried the following report.

Factory production test pilot Bob Buettgenbach walked out to the gleaming new Baron and gave me a personalized cockpit check. Our preflight disclosed that the right cowl flap wouldn't close, and we waited untl after the lunch break so that two mechanics could check the linkage. That done, I was ready to fly.

Fig. 4-14. A new Beech Baron for Van Nuys, CA, certainly beats waiting for an occasional airline. Dave Cotten gives me the word at the Beech Delivery Center.

Fig. 4-15. The speedy Beech Baron races across the western plains toward Van Nuys, CA (courtesy Bob Beuttenbach).

With both the Baron's fans spinning comfortably, I taxied to the runup area, went through the printed checklist, turned on the nav/com equipment, waggled the controls for a second time, and taxied back to the takeoff area. Beech Tower gave the nod and N8785R and I were off and running.

With one occupant, minimum baggage, and extended-range fuel tanks providing 136 gallons usable (142 total), the Baron's takeoff performance was spectacular. I had to come way back on the power halfway through my first turn to remain below the busy jet pattern at nearby McConnell Air Force Base.

Once clear of the control zone, I opened my flight plan with Wichita Radio, re-scanned the engine instrument panel at the right of the cockpit, and started a cruise-climb, pulling 2400 rpm and 24 inches of manifold pressure (Fig. 4-15).

Winds were reported light and variable, and I leveled off at 8500 feet. The right cowl flap still didn't close completely, but the gauges showed little or no difference in head temperatures, exhaust gas temperatures, or oil cooling, so I could see no valid reason to turn back.

There's really only one set of controls that's ever given me a moment's question on any of Beech's post-World War II twins. Beech is the only twin manufacturer I know of that installs the power levers with the props on the left of the console and the

Fig. 4-16. Throttle quadrant arrangement on the Beech Baron is not standard, as compared with most twins. In this model of the Beech, prop controls are on the left, throttles in the middle and mixtures on the right.

throttles in the middle. Mixture controls are standard, to the far right of the console. Frankly, during my delivery flight west and during the slow-flight exercise, I didn't touch a single lever on the B55 console without taking a second look to be doubly sure I had the control I wanted (Fig. 4-16).

"Just remember to reach for a gauge'," advised the factory, and this approach does the job very well as you survey the Baron's instrument panel. The dual tachometer is on the left, directly above the prop control; the dual manifold pressure gauge is directly above the center-mounted throttles; and the dual fuel-flow gauge is directly above the mixture controls.

I passed over Liberal, KS, at 1:43 p.m. CST. With 2300 rpm and 22 inches mp (63 percent power) and an outside air temperature of 45 degrees, my true airspeed was 205 mph. According to the fuel-flow meters. I was pulling nine gph per engine.

The checkpoints went by rapidly; Guymon, OK, at 1:58; Dalhart, TX, at 2:20; Anton Chico, NM, at 3:06. At 3:43 p.m., Eight Five Romeo and I eased down to a smooth, but not-too-short, landing at Albuquerque International—550 statute miles in two hours and 50 minutes, including off-course turns at both takeoff and landing.

Cutter Flying Service pumped 88 gallons of fuel into the Baron's tanks, since I'd been running the new powerplants just a little on the rich side. Two of Cutter's mechanics took a look at the right cowl flaps and worked on them with only slightly better results than the factory mechanics had gotten.

The local FSS was reporting "severe clear," and I filed VFR on V-12 to Van Nuys. With an en route time estimate of three hours and 30 minutes, I had wheels up at 4:57 p.m. CST.

The forecast held, the winds were light, and I climbed to 10,500 feet to clear Lookout Mountain and El Morro National Monument, east of Zuni, NM.

Having made deliveries of smaller, slower craft over this basic route for more than 25 years, I found it almost unreal to watch the well-known checkpoints pop up so quickly. At 10,500 feet (OAT 50 degrees F.), with 2300 rpm and 21 ½ inches mp (that's full throttle), I was indicating 182 mph, which trued-out at a whopping 220.

Westbound out of Albuquerque, I had an added nav—aid that I had enjoyed on other recent trips. Air carrier jet contrails strung out above and ahead of me, as the big birds and their passengers

Fig.4-17. A late afternoon sun sinks over the Colorado River as a Beech Baron speeds toward an evening landing at Van Nuys. Most ferry flights have beautiful moments like this.

headed for California, making transient signposts in the technicolor of an early evening sky.

The Baron's Mitchell Century IV autopilot didn't stay on for long, since 85R was plain fun to hand-fly. The 260-hp Continentals purred contentedly and the Hartzell prop governors held near-perfect "sync." Zuni slid tailward at 5:43 p.m.; Holbrook, AZ, at 5:59. Flagstaff and Humphrey's Peak zapped past the right wing at 6:18; Prescott at 6:36, and Needles, CA, at 7:09 (Fig. 4-17).

A chamber of commerce sunset was developing over the nose of the Baron as I checked in with Daggett at 7:44 p.m. Abeam Palmdale, I canceled VFR.

A quick change of frequencies gave me the Van Nuys ATIS, which was calling for landing on Runway 16 Left. Inbound to Newhall Pass, I called the tower as the big-city lights began to twinkle brightly in the dusk.

Both landing lights came on as I entered the busy San Fernando Valley and slowed to 175 mph (emergency wheels-down speed is 200 mph). As the gear came down, I made a careful review of the pre-landing checklist.

"Van Nuys Tower, Baron Eight Five Romeo abeam west with a light."

"Roger, Eight Five Romeo. Follow the trainer turning final. Cleared to land 16 Left."

The tires yelped briefly and the nose gear touched down at 8:17 p.m. CST—an even three hours and 20 minutes from Albuquerque, including 180-degree turns at both ends of the trip.

Fig. 4-18. Piper's factory Aztec takes on fuel at Tampico, Mexico, during the filming of a travelogue called "Tigers of the Sea." Almost all the flying was strictly ferrying.

When the tanks were filled the next day, 85R took 91 gallons, so I was still running on the rich side. An adjustment of the fuel-flow meters might have been in line, since in-flight photos showed the meters indicating eight to nine gph per engine.

When I chopped the mixture in front of Beech West and looked at the deepening sunset, I couldn't suppress a smile; I was an hour and a half ahead of the scheduled airliner I'd canceled after lunch.

Twin to Baja

When you contract with a flying service or a factory for a flying assignment, much of the time can be spent in ferrying. A good example was a month-long job for Piper Aircraft to Baja California. The aircraft factory was looking for a twin-engine pilot with some travel film experience who could do a 30-day consulting job. I had the time available and the money was right. This assignment, of course, required a commercial license with multi-engine and instrument ratings. Previously, I had produced two simple travelogues for Piper entitled *Wings to Alaska* and *Wings to Baja.*

A travelogue entitled *Tigers of the Sea* was to be filmed on the mainland of Mexico and on the Baja California peninsula. The factory was swapping the transportation capability of their twin in exchange for considerable promotional exposure in the film. After an extensive check-out at Lock Haven, PA, by factory pilot Bob Yoder, I picked up Aztec C N57324 and headed for Mexico (Fig. 4-18). This particular check ride covered everything from wet IFR through the full single-engine performance envelope and was probably much more extensive than a routine customer flight check because the factory owned the airplane!

I picked up three members of the production crew in Dallas, complete with heavy baggage and cinematographic gear; then went on to Tampico, Mexico, where three vans from the film company and the staff had congregated. After a couple of days shooting on the gulf coast of Mexico, the location shifted to the top of Baja California—a long, dreary drive and ferry trip for the vans, but just a one-stop fuel hop in the Aztec C.

A part of this flying assignment included picking up VIP's for the film company in Mexico City and shuttling them back to the tip of Baja. You had to be just a little vague about your assignment here because we were really ferrying passengers without pay in competition with the Mexican scheduled airline. On one of these shuttle trips, we had two executives from Penn Reels, a fishing tackle company in the northeast. My two passengers had never before been in Mexico and their first view off the airliner was our landing for fuel at Guadalajara. Then on we went across 200 miles of water for a landing at the relatively primitive gravel flight strip at Rancho Buena Vista. I circled the rustic resort and pointed it out. Neither passenger could recognize the "airport" until we were on short final approach, but the Aztec handles such contingencies with "no sweat." (See Fig. 4-19.).

Fig. 4-19. Cinematographers Bob Dienthal, left, and Joe Domin load their heavy camera equipment aboard the Aztec C at a dirt flight strip called Rancho Buena Vista near the tip of Baja California, ' Mexico.

Fig. 4-20. Twin-engine aircraft will carry quite a load, as shown here at Miegs Field, Chicago, where two cinematographers deplane from a Piper factory Aztec C. On occasion, twins can be almost twice the fun.

There's a certain informality in assignments of this type. When one plane load was to go from Baja's tip back to Mexico City, we decided that it might be fun to visit Puerta Vallarta for the night, since none of my passengers had ever been there. No sooner said than done.

While not ferrying and waiting, the film script called for footage of people getting in and out of the Aztec; start-up; takeoff from these rustic flight strips; and some air-to-air footage with Chicago-based cinematographers, Bob Dienthal and Joe Domin, riding in a local Cessna 206 (Fig. 4-20). At the conclusion of the filming, I ferried the two cameramen to Chicago with an overnight stop in Los Angeles—it really isn't that far out of the way, is it?

Total flight time for the month was 73.7 hours. Time on the screen of the travel film was well under five minutes. And would you believe that I couldn't find a new airplane out of Lock Haven to ferry even part way to the West Coast! Piper covered the expenses and I came west on the big tin bird.

Chapter 5

How It All Began

The largest aircraft ferrying organization yet established was called the "Army Air Corps Ferry Command" and was established prior to Pearl Harbor on May 29, 1941, with two officers and a civilian secretary. Their charter was "to move aircraft by air from factories to such terminals as may be designated by the Chief of the Army Air Corps."

Originally, the effort was to move British aircraft to Canada for shipboard crating of the smaller planes and North Atlantic ferry flights by British pilots of the larger twins.

Thirteen months later, on June 30, 1942, the unit had mushroomed to over 11,000 officers and enlisted men. They had delivered 13,595 aircraft before the name was changed to the Ferry Division of the Air Transport Command. The Ferry Division promptly grew to 125,000 people, 80,000 of whom were stationed overseas. Ballpark statistics indicate that 30,000 planes were ferried in 1942, 70,000 in 1943, 108,000 in 1944, and another 57,000 before V-J Day.

A small, but well-known, part of the Ferry Command was the 1,830 W.A.S.P.S. (Women's Air Force Service Pilots) who not only did ferrying, but some of the more unpleasant flying jobs like target towing, simulated strafing, smoke-laying, radar-jamming and searchlight tracking. One of the printable limericks about the WASPS could be sung to the tune of "Bell Bottom Trousers":

"Zoot suits and parachutes
And wings of silver, too

He'll ferry planes
Like his mama used to do."

My Gooney Bird Goes Down Under

Old log books, normally kept now for IRS verification, show that I was involved in perhaps a dozen of these deliveries in 1944. The aircraft ranged from AT-6's to B-25's, one co-pilot trip in a B-17 and several C-47 (DC-3) deliveries. The longest and most interesting of these ferry flights, before being transferred to Chabua, Assam, India, and a tour with C-46's on the Burma Hump, was the assignment to deliver a new C-47 (No. 43-16206) from Long Beach to Townsville, Australia. The C-47, not noted for its speed at economy cruise, took 52½ hours, but this was the best trip I ever had in a military airplane (Fig. 5-1).

Among the then-secret orders cut by the 6th Ferrying Group, Ferrying Division, Air Transport Command, Long Beach Army Air Field on 25 July, 1944, was one that stated, "the following personnel will proceed at proper time from Long Beach to Townsville, Australia, for the purposes of delivering aircraft.

Crew No. 2 F/O Donald C. Downie (2223706) 14th Sq. A C Pilot, F/O James P. Hanley (T-224129) 52nd Sq. A C CoPilot. Also included in the orders were a navigator, engineer and radio operator. Aircraft and crews of the ATC engaged in the operation of air transportation and ferrying services will not be delayed in such operations by theater, or similar commanders unless conditions make such delay necessary for the security of ATC aircraft and crews.

At the time these orders were cut, I had a total of 37 hours logged as pilot-in-command of twin-engined aircraft, although my total flight time with tow planes and Stearman trainers was high—2400 hours.

We logged an hour and a half of shake-down time and crew coordination before an early evening weather briefing at Long Beach. Weather was no problem for any of the six C-47's that took off that evening. Our gross weight with eight 100-gallon "Tokyo Tanks" filled in the cabin was 31,500 pounds. The airlines flew the same aircraft at 25,500 pounds. (Fig. 5-2).

After briefing, we threw our B-4 bags and parachutes in the back of No. 43-16206, climbed aboard, did a by-the-book checkout, and were soon in the air. With both Pratt & Whitney 1200-hp engines turning smoothly, our overloaded takeoff was less than 3000 feet and we climbed slowly toward 8000 feet.

When information is supplied in the blanks hereon this form becomes CONFIDENTIAL and will be handled in accordance with AR 380-5.

HEADQUARTERS, STATION #5
PACIFIC WING, AIR TRANSPORT COMMAND
ARMY AIR FORCES

28 Jul 44
(Date)

SUBJECT: Travel Orders.

TO : F/O DONALD C. DOWNIE, T 223 706

Under authority of AAF Reg 55-14 dtd 29 Mar 43 and ltr, 300.4, Hq, PW-ATC, to CO, Sta #5, PW-ATC, Sub: "Authority to Issue Travel Orders and Instructions Governing", 23 May 44, you are directed to proceed with your crew in

C-47B 43-16206 (Airplane Type & Number) to Townsville, Australia (Destination) (5th AF). . . (Air Force)

via Christmas Island; [illegible] Tutuila, Samoa Islands; Nandi, Fiji Islands; Guadalcanal, Solomon Islands; Port Moresby, New Guinea. . .
(Stops)

By order of Lieutenant Colonel WEHLE:

MARSHALL A. ELKINS
Lt Col., Air Corps
Operations Officer

OFFICIAL:

Henry W. Beardsley

HENRY W. BEARDSLEY
Captain, Air Corps
Asst Operations Officer

Fig. 5-1. Delivery orders during WW II for a C-47B from Long Beach, CA, to Townsville, Australia. The rank of F/O (Flight Officer) with a blue and gold bar was similar to the Warrant Officer rating.

Back on the ground, Ruth talked with the Weather Briefing officer who assured my wife that the C-47 had an excellent record and "would usually float for eight days if ditched properly."

It was a long, uneventful night. I'd made the first part of this trip in the co-pilot's seat of a B-25J just three weeks earlier and had landed at Hilo 21 hours and 40 minutes later with every tank in the airplane dry except the 100-gallon radio operator's tank. The slower C-47B took 15½ hours, but we still had over 400 gallons of wonderful fuel aboard when Diamond Head came over the horizon.

This long first leg wasn't entirely uneventful. We were perhaps an hour out, with no land yet in sight, when Jim Hanley nudged me from the right seat and pointed out his window. What do you know—there was a U.S. Marine Corsair flying off our wing. I looked out my window and saw another blue fighter come alongside. Soon there were three fighters on each wingtip, or closer. A radio blackout was in effect, so we didn't pick up the mike and say "howdy." The flight leader on my side slid back his canopy and pointed down with his thumb. I nodded and eased back on the power, dropping from 8000 to level off at 1000 feet over the blue Pacific.

The fighters just hung in there as we puzzled on why. As Diamond Head showed, we made the required 360-degree recognition circle shown in that day's flight log and headed for Hickam Field and the adjoining John Rogers Airport. The fighters stayed with us until our landing gear came down. Then they peeled off for parts unknown and we drifted in for that nice feeling of solid coral after so many hours of water.

We taxied in and shut down, still wondering about the escort. The cabin door was hardly open when a jeep came roaring up driven by a flustered Captain who asked briskly for my radio operator. Sgt. Mays replied, both of them headed for the aft of the airport and then took off for the flight line. Only later did we find out that another of our six-plane unit had taken off from Long Beach with its IFF (Identification Friend or Foe)—then a super-secret electronic box with a red "destruct" button on it in case of capture—safetied on emergency. This unit, similar in operation to today's transponder, had been squawking "emergency" for almost 2400 miles. The Corsairs had flown out to the first C-47 they found and escorted it back.

The next morning, we had a briefing for Christmas Island. Very early in the Ferrying Division's activities, a camera crew had been sent out to photograph each of the islands we flew from in the South Pacific. This footage, showing the outline of the island and how the runways looked on final approach, was used as an effective briefing tool. We were cautioned to be particularly cautious about cloud shadows on the ocean that could look exactly like islands, particularly if you were hunting for one in that particular area.

Flight operations had removed four of our long-range tanks and added 2500 pounds of cargo. The trailing rubber lines to the remaining "Tokyo Tanks" still led to the back door. This was our only way of dumping fuel from those tanks. We didn't have to try it out.

A mimeographed form signed by the Assistant Operations Officer at Honolulu to all pilots concerned cautioned: "In order that

Fig. 5-2. U. S. Army Air Corps C-47's parked on the military ramp at the Long Beach, CA, factory. Since flight crews were not permitted to carry their own cameras, because of security restrictions, this report is sparsely illustrated (courtesy McDonnell Douglas).

all pilots will be familiar with call signs, destinations, frequencies, etc., of other aircraft departing this station, the following is set forth. *Look it over carefully. It may come in very handy*!" Then followed a list of the 14 aircraft headed south that day. All were headed for Christmas Island except for one fighter going to Canton Island. Nine C-47's were scheduled off at five-minute intervals. Our primary communications frequency was 4595; this in days before the advent of static-free VHF (Fig. 5-3).

Eight hours and 10 minutes later, we had wheels down on the coral strip at Christmas Island. Tall coconut palms waved in tropic breezes and already we were a long way from "Stateside." That night we watched a movie with locally-based personnel—a movie set up in the open with palm trunks for headrests. "Christmas" was a little island with a small station used primarily for ferry division flight crews.

The next day, it was another eight hours and eight minutes to Tutwilla, a hurriedly-built dirt landing strip stretching out into a beautiful lagoon in the harbor at Pago Pago.

Somewhere between Christmas Island and Pago Pago, Hanley and I were lounging in the equatorial sunshine when we were both doused with a cupfull of water. It seems that this was the formal initiation for the first crossing of the equator.

Fig. 5-3. U. S. Army Air Corps C-47B in flight near the coastline. This is similar to the model ferried to Australia by the author (courtesy McDonnell Douglas).

Much over-water navigating in the mid-Pacific is over clouds so that wind drift sighting on the surface of the water isn't possible. We were perhaps an hour from landing at Tutwilla, according to our flight plan, when a big hole opened up so that the water was visible. It seemed like a good idea to spiral down rather than make an en route letdown on instruments, and that turned out to be an excellent hunch. Once below the cloud deck, we leveled out on course, and what do you know—there was a volcanic mountain peak jutting up into the cloud base that we could have augered into on a normal approach. We'd picked up a hefty tailwind without knowing it!

The Tutwilla Airport was long and narrow. It was so narrow that no taxiway existed, so we battled a 20-knot crosswind, planted the docile C-47 on the ground and rolled to a warm-up area at the lagoon end to watch two other C-47's in our flight battle the winds down through the wide-enough cut in the jungle on final approach, touch smoothly on the upwind wheel and roll up to join us. In situations like this, you soon found out why the C-47 and its civilian counterpart, DC-3, have been referred to as the one best airplane ever built.

During the stresses and secrecy of wartime, flight crews were not permitted personal cameras on many flights. Therefore, there are no on-site illustrations of what was a beautiful, exciting trip.

Another morning briefing was held on the airport at Nandi in Fiji. Film clips showed a hump in the middle of the runway and the bored briefing officer mentioned that a number of tail-wheel transports like our C-47 had been nosed-up by crews who thought they were running out of runway as the crown of the airport came closer. As I remember it, there were 3000 feet of additional runway over the hill and usable. The daylight flight was 10 minutes under six hours.

Another short night on the "town" and we had a chance to purchase Indian filigree from dark-skinned merchants who looked at their uniformed visitors as gullible souvenier hunters. They were right.

Next stop, 6:40 to Guadalcanal and the war seemed very close. Enlisted men in the crew advised that transient quarters were less than satisfactory, and I tried unsuccessfully to have our entire five-man crew billeted in the transient officers' quarters. When that didn't work, we all decided to sleep on the cargo in the airplane, and no one on the island objected.

A borrowed jeep gave us a look at the long lines of military graves, many with the same closing date when a large ammunition dump had exploded. For those of us on our first trip south, it was a very sobering experience.

The new C-47 was remarkably free of "bugs," and we were off early the next morning for Port Morsby in New Guinea. The briefing film showed an approach over jungle and a spongy steel matt runway over very wet ground. Midway over the Coral Sea, we saw a submarine surfaced, chugging along and minding its own business. One of our crew wanted to go down and buzz the sub; but I figured that they might shoot first, even at our "Gooney Bird," and make sure later if we were in fact friends. We left the transport on "George", the auto pilot, and continued along our pre-set course.

At Morsby, we topped our tanks, had a lunch of Spam sandwiches, and headed out the remaining four hours to Townsville. Lt. Coles, our navigator, was "sweating out" a new baby in Long Beach and was eager to get back. We tried to get him relieved to return from Morsby, but the Operations Officer wouldn't buy the idea—even though it would have been quite difficult to miss the continent of Australia on a four-hour flight.

Again on takeoff from Port Morsby, the war that had been going on for so long seemed very close to home. Off the departure end of the steel-mat runway was a series of wrecked airplanes—B-17's, B-25's, C-47's and several others—evidence of aircraft with full loads and probably subsequent engine failures. Fortunately, we were only at gross weight and both those ever-lovin' engines kept right on turning.

Townsville was just another ramshackle Army Air Base a distance out of town. The local Operations Officer who signed for our ship looked at my blue-and-gold Flight Officer bars and commented with some disdain that, "They're letting anyone bring 'em down here these days."

In reply, I said, "Well, you've got your airplane, haven't you, sir?" I hoped that he read the "sir" the way I meant it.

There was time for a brief dinner, a slug of drinks with a flying buddy from glider-towing days in California, and I was wrapped up on the floor of a DC-4 headed back for Morsby, Guadalcanal, Tarawa, Canon and Hickam. Tarawa, with its beaches still piled high with smashed landing craft and coconut palms shredded by shell fire, looked stark and grim under a full moon. Then it was 13 hours in a lumbering LB-30, the passenger-carrying version of the B-24, back in to Hamilton Field north of San Francisco. The return trip, including stops, was two days and eight hours. All of a sudden, the world had shrunk, for me, to fit the speeds of the air age.

It shrank again when the passenger jets came in.

Jet to the Far North

Ferrying military aircraft has one great advantage. There's a large, powerful, and usually, extremely efficient team for back-up. The same sort of support found in the Pacific during WW II was also available many years later to the military pilots of jet fighters going from the factory to Alaska.

Delivery of military jets has changed over the years. Long over-water refueling flights are now a tactical routine that delivers fighter-type aircraft to virtually any spot in the world. In the days before in-flight refueling was an accepted procedure, military deliveries were just about the same as civil lightplane trips with a landing at a big airport every couple of hours.

Back in the days when temperatures and tempers were warm along the western Alaskan shoreline, our first line of defense was the radar-equipped Lockheed F-94B. The planes were manufactured in Los Angeles and many were ferried to Alaska by the fighter pilots who would use them. For security reasons, original publication of the following article (in the Pasadena *Star News* and "Skyways" magazine) was delayed by the Department of Defense until the changeover from F-80's to F-94's was completed.

No one was more surprised than the flight-line alert crew at Lockheed's Van Nuys, CA, factory to see a correspondent, complete with Speed Graphic and lightweight portable typewriter, unload his bag into the nose of a northbound fighter and then climb into the cockpit of the '94.

Aside from the ammunition cans that were removed from the nose section to make room for a skimpy supply of luggage, there was no modification made for this ferry flight. The six planes in this flight were all standard-equipped with 230-gallon Fletcher tip-tanks and were ferried with their complete installation of guns and radar equipment.

"We'll fly daylight only and stay out of as much weather as possible," explained Captain Conrad Johnson of Fergus Falls, MN. He drew the job of carrying the correspondent in the back seat of his airplane. He's 33 years old, has over 3000 hours of single-engine time and has been flying jets for over four years (Fig. 5-4).

"I'll check you out on the emergency equipment in the plane," said Captain Johnson as we walked out toward the big red-tailed fighter. "First, you don't pull any emergency control unless I tell you to or I've obviously passed out. In that case the yellow handle here on the right will jettison the canopy. There's one of these handles in each cockpit to fire the explosive charge that will blow off the plexiglas. Then you arm the ejection seat by pulling up on these two arm rests. Make sure that both feet are on the stirrups before you pull the trigger. That'll blow you clear of the airplane. Then pop your safety belt to get away from the seat and open your 'chute normally.

"I doubt that you'll have any occasion to use that equipment," he added with a grin, "But since it's in the airplane, you'd better know how it operates."

Then we wrestled into a back-pack parachute and crash helmet with oxygen mask, and climbed into the back seat. Captain Johnson helped hook up the oxygen and showed us how the shoulder straps hooked into the safety belt. When you're tied into one of these fighters, you're really there to stay.

The pilot climbed into the front cockpit and quickly donned his own safety equipment. The yellow starter unit parked alongside burst into power and soon the whine of our own turbojet engine was heard.

Fig. 5-4. Captain Conrad Johnson of Fergus Falls, MN, grins out of the front seat of a USAF F-94B on the ramp of the Lockheed Delivery Center in Van Nuys, CA. Pilot in the back seat hiding behind an oxygen mask is the author.

"Pull in your arms," said Captain Johnson over the intercom. "The canopy is coming down now." The electric motor closed the big clamshell quickly, pressure in the cabin increased for an instant and then went back to normal. If you have any tendency toward claustrophobia, stay out of the back seat of one of these new fighters.

The intercom system is switched on between both cockpits all the time, so you don't have to mash down on any button to talk to the pilot. In addition, the back-seat passenger can hear all radio transmission on the VHF channels.

With two other F-94's flown by Captain Roy Barnes of San Francisco and Captain Robert T. Goetz of Osceola, AK, we taxied away from the factory acceptance line. The three other ships of this flight had departed for Las Vegas, NV, the night before.

We lined up at the end of the runway in tight formation and started to roll. "There's no complicated warm-up," explained the pilot as we started our roll. "Just make sure that your fuel pumps and electrical system are working. Then walk forward on the throttle and you're in business."

As our speed picked up, Captain Goetz, the flight leader, said, "After-burner—now." You could feel the surge as our speed picked up even faster. The after-burner is a blow torch-type fuel-burning system that gives additional thrust for takeoff and short periods of emergency climb.

Our takeoff was made on a hot mid-afternoon and the jets took most of the runway to break ground. As the ship broke ground and started to climb, I asked Captain Barnes about the length of run.

"No sweat," said the pilot easily. "We had plenty of room and nearly 160 knots (185 mph) over the trees. There are a couple of fields up the line that are tighter than that," he added cheerfully.

Standard procedure for these fighter pilots is to take off and climb in close formation (Fig. 5-5). The reason for this is partly essential practice for the boys' tactical efficiency, and partly "good show" for the taxpayers. And these boys fly a "good show." From this close range, the F-94's look like a king-sized Coca-Cola bottle with stubby wings added as an afterthought—and you don't have to strain your eyes to read any of the small print on the other planes. The three ships circled the San Fernando valley and began climbing on course. The altimeter hand spun like a fast-running second hand on a watch and we leveled off over Daggett at cruising speed—somewhere around seven miles per minute.

"Navigation in these jets isn't too much different from anything else," explained Captain Johnson as we sped toward Las Vegas. "The main difference is that things happen faster. The new jet charts—six cover the whole United States—are nice, but all of us carry regular old-fashioned strip maps in addition to a radio-facilities chart.

"We normally fly airways in the States—and *always* airways through Canada and Alaska. Between the range stations and the radio compass, we can keep an accurate check of our position, no matter the speed. About the only time that any of us get into trouble is over our home station doing lazy eights or acrobatics at the end of a flight to burn out enough fuel to land comfortably. It is quite easy to drift 30 or 40 miles away from home in just a few minutes. Then you're in a 'bind' because your fuel is about gone. But on a regular cross-country flight, we're never 10 miles away from the airway.

"Watch along the way and see how close we hit our estimated time of arrival. If we're more than two-and-a-half or three minutes off, someone went to sleep."

The tight formation was relaxed and the pilots settled down to follow the beam. In cross-country navigation of this type, the flight leader actually does the navigating and makes the radio contacts. However, each pilot has a full set of maps and follows the progress of the flight very closely so that he can take off on his own should he become lost from the other planes.

There was absolutely no sensation of speed until the weather clouded over and we pulled back into close formation. Then, as the wisps of clouds began to flick by, you really knew you were travel-

Fig. 5-5. Formation takeoffs are routine for military ferry pilots. Here an F-94B breaks ground at Van Nuys while the second plane serves as a camera platform.

ing. There was no vibration, very little noise and the three ships seemed to hang in the air on invisible strings, controlled by a smooth-handed puppeteer.

No sooner did the three planes form and start climbing in formation than we ran through a thin, wispy curtain of clouds into clear air again.

The Great Salt Lake of Utah rolled into sight and we circled in for a landing at Hill Field in Ogden (Fig. 5-6).

"These F-94's are quite heavy, and they land 'hot'," said Captain Johnson over the intercom. "We'll probably touch down at about 110 knots (126 mph) at this altitude."

The scenery was streaming by at a terrific clip when the wheels lightly touched the runway.

"This is strictly a brake airplane," said the pilot. "It takes plenty of brake pressure to slow down at all. The average transport-plane runway is barely big enough for a jet."

It was wonderful to get out and stretch. After I complained a bit about being cramped in the back cockpit, Lt. Edward Wynuck of Detroit found me a rubber cushion for the remainder of the trip. I really believe it saved me from a fate worse than death.

At Ogden we joined the other three Alaska-bound interceptors. Their pilots were Captain Joe Wisby of Springfield, MO and Lt. Wynuck and Lt. Felix A. "Doc" Blanchard of McColl, South Carolina. Sports fans will remember Lt. Blanchard as "Mr. Inside" of the famed Army football team of 1945-46.

While the jets were refueled, the pilots took on coffee and sandwiches. It wasn't long before we were strapped back in the planes and taxiing out for takeoff. The air was cool and "heavy" at

Ogden, so the planes broke ground a comfortable distance from the end of the runway.

"No sweat," said Captain Johnson cheerfully as we circled out over the lake and headed on course. Flecks of ice streamed in through the ventilators. "When the moisture content of the air is high, we frequently pick of this type of ice from our own cockpit refrigeration unit," explained the pilot.

"Ice knocks off the paint, and a heavy rain storm will do that, too. From the looks of this cloud deck up ahead, I'll have a chance to show you," he continued.

These Alaskan jets all carry red outer-wing panels and red tail surfaces that will show up well against the whiteness of the Arctic in case of a forced landing.

There was a quick minute of instruments and then we broke out in the clear and started to descend toward Great ' Falls. Even though our take-off time had been nearly three in the afternoon, we landed here in northern Idaho long before dark. Ironically, our route was similar to the old Ferry Division airway that took fighters to Russia during World War II.

It took Captain Johnson seven days—and three blow outs—to drive his car from Great Falls to Anchorage. On his first delivery flight, he made the whole trip in a leisurely day. We started off with six F-94's the next morning to repeat the process, but didn't quite make it.

During a routine post-flight inspection at Edmonton, Canada, a faint burned spot was noticed on the tail pipe of Captain Barnes' plane. Further investigation revealed a slightly cracked weld that, though not dangerous, should be replaced. By the simple process of matching pennies, the pilots decided who would stay and who would continue. Among the pilots who remained was my aerial driver.

We never fly around this country alone," explained Captain Barnes as we sat around the Canadian ready room. "These airplanes cost roughly $300,000 each. There's no reason to lose one, or have a pilot wind up missing in the wild country ahead just because he was flying alone. Our planes always go out in pairs at least unless they're in the immediate vicinity of their own home field.

"We'll have to wait here for parts to be flown down from Alaska or up from the factory," said Captain Barnes. "The 'Gooney Bird' has all our spare tires and minor parts, but it doesn't have a new tail pipe."

Fig. 5-6. Flight of F-94B's bound for Alaska over snowcapped peaks near Ogden, Utah.

The "Gooney Bird" is the Douglas C-47 that acts as a mother ship for these six jets. Each jet, with the exception of Captain Johnson's, had it's own crew chief in the back cockpit. These flying mechanics had attended school at Lockheed before returning with their new airplanes.

While the teletype circuit was humming with the request for parts, the flight mechanics removed the old tail pipe and readied the F-94 for a quick change.

The juke box in the flight-line coffee shop at Edmonton had two selections popular in Alaska and Canada, but not yet on the hit parade in the States. Their titles were "When the Ice Worms Nest Again" and "The Squaws Along the Yukon Are Good Enough for Me." "Doc" Blanchard put in a nickel and played "Way Down Upon the Swanee River."

After the repairs were completed, our three planes took off from Edmonton with Captain Wisby on the right wing. We climbed to 25,000 feet, cruised on top over Whitecourt and Grand Prairie Radio. A scattered undercast changed slowly into solid clouds over St. John and we cruised along in the bright-blue Canadian sky on top. Over Beatton River Radio, Captain Barnes called the Air Traffic Control for permission to make a penetration let down into Fort Nelson where the current weather was reported as 9,000 feet with lower scattered clouds and 30-mile visibility (Fig. 5-7).

The three ships bunched together like baby chickens burrowing under a mother hen, and Captain Barnes called "Reduce to 85 percent; dive flaps—now."

The F-94's shuddered slightly as the dive flaps took hold. Then the lead plane stuck its nose into the clouds and we started down. The let down was smooth and uneventful and the three jets popped out below the overcast about four minutes before they reached the station.

Fort Nelson is a marginal airport along the route for those heavy jets. Since the F-94 wings are thin and the bulk of the fuselage is loaded with engine, radar equipment and the two-man crew, there isn't too much space to put the landing gear. Consequently, rather small, high-pressure tires are used. Average jet-airplane tire life is short—even on long runways where excessive braking is not necessary.

Our three planes circled under the overcast for about 10 minutes to burn out their remaining fuel, and we landed with less than 100 gallons aboard. No one ran off the end of the runway and the pilots wise-cracked among themselves as they taxied up to the line.

On our next hop, I had a chance to see, first hand, why the F-94 is called an "all-weather fighter." Weather reported the forecast for Whitehorse as 8000 feet with lower scattered clouds, so we fired up the planes and took off. On this leg of the flight, Captain Wisby was leading.

After takeoff, we circled the rain-soaked airport and headed out on course, climbing up through 20,000 feet of solid overcast (Fig. 5-8). Rain streaked back over the plexiglas canopy in pencil-thin lines and showed in horizontal streaks past the wing tips. You could see the paint disappear from the red leading edges as the three planes plowed upward through the wet weather.

Fig. 5-7. An F-94B "slugs" fuel from a vent. This is normal and stops after a few seconds.

Fig. 5-8. Full formation of six F-94B's en route to Alaska. At the time this trip was made, all landing places were classified and publication of the resulting report was delayed for six months by the U. S. Department of Defense.

Just as we broke out on top and loosened the formation, the Fort Nelson radio called, "Jet zero nine zero six, special weather report. Whitehorse has just dropped from 8000 to 5000 feet. The forecaster advises that you return to base."

Captain Wisby called back immediately, "Will you get me the latest Northway weather. Northway was the next airport north of Whitehorse. Let's go to 97 percent and get some altitude."

The higher the altitude, the more efficient are these jet engines.

Fort Nelson called again within a minute. "Whitehorse is still 5000 but deteriorating rapidly. The forecaster still advises that you return."

"Starting left turn and reduce power to 90 percent," said Captain Wisby. "Nelson tower from jet zero nine zero six, what is your latest weather and advise GCA that we are returning."

"The current Fort Nelson weather is 1600 feet overcast with lower broken clouds, light rain and five-mile visibility. GCA is standing by."

Tech. Sgt. Roland Cooper of Hamilton, OH, riding in the back seat of Captain Wisby's plane, put his head down and flexed his big hands. "I'm really going to sweat this one out," he told the pilot over the intercom.

From 25,000 feet on down, there was little to do but watch the two other planes bob around slightly in formation and listen to the talk on the radio.

"Reading you loud and clear on GCA," said the ground station, "but you are not yet on the radar scope."

"Did you know that you just passed Fort Nelson?" asked Captain Johnson. The planes were in such close formation that the flight leader merely nodded. On an actual letdown of this type, the flight leader does all the navigation and makes the radio contacts.

"Do you have us in the scope yet?" asked Captain Wisby.

"Negative," was the reply. "You're still too high. What is your present position?"

"We're headed inbound to the station on the north leg of the range. Present altitude 13,000 feet," answered the flight leader.

"I've got a touch of vertigo," said my pilot over the intercom. "When you're flying instruments like this and watching the lead ship so closely, vertigo happens quite frequently. You can always tell when I'm having that sensation. I fly about six feet higher than normal in formation."

"We are now inbound descending to 11,000," said Captain Wisby. "Let's up the power to 90 percent and it'll be easier to fly formation—now."

This was the real McCoy. Normally in the States, a letdown of this type is simulated for pilot practice, but not today. We were making a normal range letdown and the GCA crew was standing by to bring us in on a low approach. This is the pay-off for hundreds of hours of practice and thousands of dollars of expense, but it all saved a million dollars worth of airplanes and, incidentally, six guys who were in no particular hurry to meet their Maker.

"We are now five miles west of the station doing a turn around," said Captain Wisby.

"We now have you in good radar contact," replied the GCA crew. "Turn right to 200 degrees for positive identification."

On solid instruments, the only way I could tell that we were turning was to watch the peel-up of the wings and feel the "G" load.

"You are now 12 miles south of the station. Continue your turn to a heading of 200 degrees. Descend to and maintain 3000 feet," said the GCA operator. "Your current weather is 1600 feet overcast with lower broken clouds."

The sickly gray of the cloud interior changed briefly to gray-green and the tops of the pine trees flicked into sight intermittently as we sped through the bottom edges of the clouds. I was wearing an old physical education "sweat suit" under my flying overalls—and sweating.

Again the forests appeared and the three jets streaked in under the weather with a slot of over a thousand feet between the cloud base and the tree tops. It was just like flying through the crack under a swinging door.

"Let's go ahead and finish the GCA run for practice," said Captain Wisby. "Spread out and drop your gear—now Tell me Whe we're one mile from the glide path."

"This is your final controller," said the GCA transmitter. "Do not acknowledge any further transmission. You are now five miles from touchdown and coming up fast on the glide path. Turn to zero two five degrees and descend 700 feet per minute. You are now 3 ½ miles from touchdown."

The changes in heading followed at one- or two-degree intervals and I leaned out to watch the other two planes on our right. Through the rounded curve of the canopy, I could see the runway looming up dead ahead. We were lined up perfectly.

"You are now over the end of the runway," said the operator in the GCA shack who could "see" us only on his landing scope.

"Good run," said Captain Wisby as the three ships sucked up their wheels and pulled up 500 feet to burn out some fuel. With precision instrument pilots like these three, there's "no sweat" about going anywhere in the world—on instruments. It was surprising, at least to me, to find such a sharp GCA crew at such an isolated location.

There's none of the "rah-rah, Into-the Wild-Blue-Yonder" about this group of fighter pilots. They're mature—all these three are over 30 years old. They all seem to look upon flying as a business.

"There isn't any room for the 'hot-rocks' in the ships we're flying," said Captain Wisby after our uneventful landing. "We're flying expensive equipment on a precision job. We don't take any chances."

"This one was routine." said Captain Barnes. "The roughest letdown I've ever made was the time I followed Col. William P. Benedict—our former Commanding Officer—into Kodiak in the rain with the ceiling down to 200 feet and a quarter of a mile visibility. We were flying F-80's, and the Colonel had 'Doc' Blanchard flying on his wing. When they broke out under the weather all lined up for an easy landing, the Colonel said, "Wheels up. Let's go around and do it again. This is good practice.' That's how we learned to fly instruments in a fighter."

"The important part about a jet letdown," explained Captain Wisby, "lies in being able to 'track' properly into the station. We have some wild winds up in Alaska that frequently go over 200 mph. If you just 'bird dog' (follow the needle of your radio compass) on the station, you can easily plow into a mountain. We teach all the new pilots who come up here from the States how to track into the station on a given heading in the T-33 before we check them out in any of the single-seat planes.

"When there are a lot of electrical disturbances in the sky we have to track into the station using our radio compass in the 'null' position—flying toward the no-sound spot in front of the antenna.

"One of the older pilots who just joined our squadron after being out of flying for a few years told me that he was just now really beginning to learn to fly. He said that the fighters he flew in the last war didn't require any of the precision flying we're doing here."

Since the Weather showed no sign of improving, we had the planes refueled and spent the night at Fort Nelson.

After our letdown into Fort Nelson, the hop on to Whitehorse was comparatively uneventful. We took off the next morning, climbed back on top and cruised for an hour into ever-improving weather. Below, snow was visible and ice remained in some of the rivers and lakes (Fig. 5-9).

Over Whitehorse, the three jets had more fuel aboard than was comfortable for a landing. A 'hot-rock' fighter pilot would have done acrobatics, but not these 'older-but-not-bolder' pilots. They practiced an instrument letdown in formation, just in case they ever had to go into Whitehorse in dirty weather.

Fig. 5-9. USAF interceptor over snowcapped peaks en route to Alaska.

Fig. 5-10. Alaskan mountains reach up toward the overcast as the flight of F-94B's nears Anchorage, AK.

While the planes were being refueled, we had a chance to go downtown and look at the fabulous river boats of the Yukon. These huge stern-wheelers draw only four feet of water and navigate the river down to Skagway. The boats are pulled out of the water during the winter and were just going back into commission. Downstream it's two days, and it's five days back up again. The woodburning engines take a log a minute on the upstream trip. Our F-94's could cruise the same distance in about 12 minutes.

The last leg "home" for the pilots was by far the most awe-inspiring of the whole trip. We flew up the airways over the radio range stations with the picturesque names of Aishihik, Snag and Northway. Snag, in Canada, is reportedly a colder spot than all Alaska during the winter months. The country below was completely isolated and even single cabins were marked on the airways maps.

From Northway, we crossed the Alaskan range and began to head for the coast. "Up ahead are the glaciers," said Captain Johnson. We had a break in the weather and only scattered clouds covered these white mountain peaks. The broken pattern of the crevasses is beautiful to view from the air, but it would be more than a minor problem were you on the ground.

"I'll bet there isn't a living soul within 50 miles of us," said Captain Johnson over the intercom. He emphasized "living" just slightly.

The whole area looked like the last reel from *Lost Horizons*, multiplied about six times and projected on a giant blue-white screen (Fig. 5-10).

"If we had a 'flame-out'—engine failure due to fuel starvation—here, I'd be tempted to try the trick that I heard about recently. A pilot in an F-80 had his powerplant fail and his wing man flew in behind him, jammed the nose of his fighter into the dead tail pipe and pushed the fighter back to his base." (See Fig. 5-11.)

We streaked by Mt. Marcus Baker and down over the Matanuska Glacier and into the fertile Matanuska Valley, scene of a mass migration movement of the depression era. From a very quick glance, the area looks prosperous.

Around the corner of Knik Arm, and there's Anchorage!

We pull in, circle the busy airport and land. Another red and-silver fighter, fresh from the factories of Southern California, now sits on the alert ramp here at Anchorage. And the way they travel, it isn't far from home.

The particular F-94B in which I flew to Alaska later had a flame-out after takeoff from Elmendorf Air Force Base. The ship "bellied-in" on a frozen lake without injury to either pilot or radar operator. However, the crew had tried to eject without results. A mechanical investigation revealed that the firing pins for the ejection seats were both too short to make contact! So, I had not been sitting on a 'hot seat' after all.

Big Tow in Wartime

A great deal of time, money and effort went into the development of military gliders during World War II. Combat cargo gliders, designed to be built from non-critical materials, were a priority project. One such development was the experimental cargo glider SCG-16 designed by Hawley Bowlus and built by the General Airborne Transport Company. First flown at Oxnard, CA, 45 miles northwest of Los Angeles, the ship was prepared for delivery to the Clinton County Army Air Corps Experimental Glider Headquarters at Wilmington, OH.

A report of this cross-country delivery flight was written for "Soaring" magazine by Harry N. Perl who preserved the accompanying illustrations taken by J. Campion. Both men worked on the SCG-16 project. Perl was one of two survivors of the subsequent crash of this prototype which killed Howard Morrison and Richard C. duPont.

The ship was thoroughly inspected, serviced, and then loaded. Ballast and a variety of items—such as spare tires for the towplane, spare towlines, ground-handling gear, and control sur-

Fig. 5-11. The Lockheed F-94B makes a striking sight from slightly aft and below the tailpipe. USAF pilots take great pride in their formation flying abilities.

face gust locks—were stowed and secured. Finally all was in readiness and the big ship was towed by jeep to the starting position at the end of the runway. The towplane, a B-17 fitted with special instrumentation and equipment for measuring power requirements and glider tow forces, was used for towing. Lt. Colonel Bruce Price, Commanding Officer of Clinton County Army Air Base, was in charge of the flight. The glider crew consisted of Hawley Bowlus, Paul Tuntland, Fred Bowlus and the author, Harry Perl.

The B-17, named "Miss Barbara," had wheeled into position, the towline—a ¾-inch-diameter nylon rope with an intercommunication line attached—was locked into position, and the intercom was tested (Fig. 5-12). All was okay, and we were ready for takeoff. The B-17 moved slowly forward and soon the turbulence from the four propellers reached the glider and peppered her with a hail of small gravel from the runway. The diameter of the towline began to diminish slowly; the glider with her 15,000-pound gross weight stubbornly remained motionless, then almost imperceptibly began to roll. As the nylon line began to return to its normal diameter, the ship accelerated smoothly forward. With the increasing speed, the glider became more buoyant and gracefully took to the air. Not so easy with the towplane, with a full load of fuel and a full complement of crew members, it trundled down the runway. The wheels lifted momentarily, only to return again to the runway. Another try and finally the wheels lifted and we were on our way.

A long, steady, straight clib took us over the town of Oxnard and out over the shores of the Pacific Ocean. At an altitude of 2000 feet, the '17' began a gradual turn to the left and then squared away on an easterly heading. The air was glassy-smooth as we continued to climb to the cruising altitude of 9000 feet. The huge, sprawling city of Los Angeles and the neighboring communities loomed ahead and slipped slowly under us. We were all on the intercom system and could hear the pilot and flight engineer in the towplane discussing power settings and other operational details. Occasionally we could see someone peering back at us from the rear gunner's position in the '17.' This compartment, incidentally, was revised for flight-test work and was equipped with special towline force meters and cameras. Everyone then settled down for the long flight to Tucson, AZ.

As we approached March Field near Riverside, CA, some 50 miles east of Los Angeles, we observed a large cloud of black smoke belching forth from the number four engine of the towplane. Immediately, over the intercom, we heard the crew of the '17' execute the emergency engine-stopping and propeller-feathering procedure. The windmilling propeller slowly ceased rotating as the blades went into the full feathering position. Colonel Price called back and instructed us to release as soon as we had sighted March Field. We already had the field in view, so away went the towline and we settled into a steady glide toward the airport. We switched over to the tower frequency and were cleared to land. Hawley Bowlus was flying. He swung the ship around, lined up the runway, and "greased her on." As we rolled to a stop, a tow car drew up and towed us off the field to the tie-down area.

Examination of the B-17 revealed that the engine had "swallowed a valve" and had a cracked piston together with a severely damaged cylinder. The repair job consumed all of the following day and after a check and run-in it was well after midnight when we were ready to resume the flight. Again we lined up at the end of the runway, and the towline was attached. We called forward to the B-17 for an okay and received an answer from the roar of the engines. The noise seemed magnified in the darkness; this was our first night takeoff. The sound of the wheels of the glider coming up to speed was eerie as we sped down the runway. Soon we were airborne and held our position until we saw the wheels of the B-17 in the reflection of the landing lights lift from the runway. The purplish hue from the engine exhaust cast an unusual glow on the tail assembly of the towplane.

The climb to cruising altitude was carried out over the field in case of emergency. Again we heard the chatter over the intercom

Fig. 5-12. B-17 tow plane "Miss Barbara" takes off with the combat cargo glider SCG-16 in tow (courtesy "Soaring Magazine" by J. Campion).

and in the background could hear the "A" and "N" signals of the radio range station as we took a heading for Tucson, AZ. The air was smooth and cold at 9000 feet. As we passed over each mountain range, we received short, sharp bits of turbulence. One bump was especially sharp, but a quick check of the cargo and ship showed all was in order. The solitude of the darkness was broken only by the whistle of the wind, the distant roar of the towplane engines, and the monotonous tone of the radio range station. The night wore on, the twinkling lights below slipped silently by and served as checkpoints for our course. We took particular pains to keep all available airports in check. The lights of Phoenix, AZ, appeared and were soon lost behind us.

Faintly, the Eastern sky began to glow and daylight was not far away. A feeling of relief was expressed by all members of the crew. It would have been no fun to come off tow over an unfriendly desert terrain during the night. As we approached Tucson, the sun was peeking over the mountains and we were cruising just under a thin stratus cloud. The condensation trails behind the propellers and the moisture leaving the trailing edge of the wing combined to produce a weird optical effect. A light rain began a tattoo on the plywood skin of the glider. We were soon out of the rain area, and the moisture disappeared quickly from the ship. Tucson slipped behind us and ahead lay a large expanse of rugged mountain terrain. Again, everyone tightened up until we had the rugged country behind us, and some turbulence was encountered over the higher ranges.

Passing over New Mexico, the sky was filled with the remnants of the previous night's thunderstorm activity. Some of these mountains of energy were still spewing forth heavy downpours. Fortunately, none lay in our path. Approaching El Paso, TX, we could see ahead to Guadalupe Pass. The pass was flanked on either side by ominous-looking, building cumulonimbus clouds. The anvil tops of these cloud structures towered above us as we slid through and took a heading for Wink, TX, our first refueling stop.

The B-17 called back and informed us that they had the field in view. We checked the position and then released from the towline. The air was smooth and comfortably cool. It was 10:30 a.m. There were no clouds visible anywhere, and we agreed it was good to have all the rough terrain behind us. As we glided down, the temperature and the humidity increased rapidly. By the time we rolled to a stop, everyone was ready to return to the cool upper altitudes.

Both ships were thoroughly checked and the B-17 was loaded with fuel and oil. The crews of both aircraft took a welcome stretch and a lunch of Texas hamburgers. Two hours were consumed in the refueling operation, and the sky had by then filled with small "cotttonball"umulus clouds. We soaring pilots naturally discussed the soaring potentialities of the SCG-16 in light of the existing conditions. We took our usual take-off positions and proceeded with the flight. As soon as we became airborne, the cause of the cumuli overhead took effect, and we realized that we were in for a rough, bumpy ride. The B-17 seemed glued to the runway and made three attempts before remaining airborne—and not at all too soon—as the runway boundary markers whizzed past. The climb was almost non-existent in the hot, humid air, and we all cast anxious eyes at the surrounding oil-well derricks, which looked much too close and far too numerous.

Slowly we began to gain altitude, but the climb to cruising level seemed an eternity and the rough air kept Hawley Bowlus at his best holding the tow position (Figs. 5-13 and 5-14). The cumulus grew quickly into the towering type, and incipient thunderstorms were forming everywhere. After two hours of the rough air, it was agreed that we should climb above the cloud level. A half an hour of climbing put us above the cloud base and into a new world of fantastic cloud forms. The smooth air was a welcome relief to everyone. We were now able to alternate on the flying. The billowy clouds cut off all but an occasional view of the earth below. Needless to say, we soaring pilots in the glider grew ecstatic over these terific soaring conditions. Occasionally we would skim near a large cumulus, with a portion of the wing tip entering

Fig. 5-13. SCG-16 in tow as seen from the tail of the B-17 "tug." (courtesy "Soaring Magazine" by J. Campion)

the cloud, and would feel the turbulence hidden within the mist. As we threaded our way along the canyons formed by the towering cloud walls, it became apparent that we would have to descend, the clouds having massed together into a solid wall.

Tulsa, OK, was directly beneath as we spiraled down to get below the cloud base. This was a tricky maneuver, but the towline never became slack—a tribute to the skill of the pilots. Again, a rough ride was the order of the day and we reluctantly reconciled ourselves to more hours of rough flying. The red-brown earth of Oklahoma soon changed to the green rolling hills of the Ozarks,

Fig. 5-14. B-17 tow plane on transcontinental ferry flight as seen from the cockpit of the SCG-16 flown by Hawley Bowlus (courtesy "Soaring Magazine" by J. Campion).

and we passed over innumerable lakes dotted with the small boats of Sunday afternoon fishermen.

A thin silver ribbon on the horizon was quickly identified as the Mississippi River, and the city of St. Louis could be seen to the northeast. The intercom broke the silence as we heard the towplane call the tower at Scott Field, IL, for landing clearance. We in the glider received the usual instructions, located the field and released from the tow. The local citizenry were agape as we circled over the town on final for landing. It was 5:30 p.m., a long day, and everyone was ready for a good meal and a soft bed.

The next morning, everyone was eager to get aloft and we were soon rolling down the runway. This was a short flight of only three hours, and we received quite a thrill as we listened to the first radio contact with the tower at the Clinton County Air Base. We received the instructions to release when we were able to locate the field. The green terrain of Ohio made it difficult to distinguish the airport. Fred Bowlus spotted the runway, the towline was released, and we were sliding down for our last landing. The electric motors actuating the landing gear whined as the gear was extended, and a small turbulence was felt as the wheels projected into the airstream. A second whine of the motors and we felt the powerful braking effect of the wing flaps, then two tire squeaks as we touched down on the runway. Hawley wheeled the ship from the landing strip to a taxiway and we rolled to a stop—within a few feet of the parking area. We were immediately greeted by the air base personnel and were escorted to the flight operations office. There we went through the formal Air Corps aircraft acceptance procedure—mission accomplished. Total flight time: 13 hours and four minutes.

Harry Perl recently supplied an epilogue to this unusual cross-country glider ferry flight. He explained:

"Due to all the delays in the program, everyone was anxious to get on with the delivery flight. Believe me, the 'pucker factor' was very high in the initial part of the flight, but we soon settled down. We had every possible landing field well marked on our charts and we checked off every one as we passed over. There were compensations, however—the very smooth air, the night view of the desert towns below and the stars above, and the spectacular welcome sunrise. It was quite a flight, and one that I will always remember."

The demise of the experimental prototype XCG-14 comes from newspaper clippings. The glider went into a spin from 3000 feet over March Field, CA, and crashed. The pilot on this last flight was Richard Chichester du Pont, then 31 and one of America's most noted glider pilots. He had won the National Soaring title five years straight at the time of the crash. Three other crew members were also killed. Harry Perl and Paul Wells parachuted safely.

Civilian Ferry Flying: Where it Came From

Ferry flying has been around a long time. In a special Piper Educational Bulletin, significantly dated December 1, 1941, here's how the "Method of Delivery" was described:

"Unlike most manufactured goods, Cubs are seldom shipped. About 85 percent of the planes manufactured are flown to their destination either by their owners or by factory ferry pilots. In 1941, the Factory Delivery System was introduced to the advantage of purchasers and employees alike. Dealers, busy with flight instruction, have been finding it increasingly difficult to spare good pilots to go to the factory for planes. At the same time, there were many good pilots working at the Piper plant anxious to make extended cross-country flights. Under the Factory Delivery System, guaranteed delivery of the planes is made by members of the Club Fliers who are paid a certain mileage rate which assures them of complete expenses. Three cents a mile is allocated for insurance on the plane.

Cubs have been flown to delivery points in many places beyond the borders of the United States. A number of Cubs have been flown to Alaska, nearly a dozen Cubs have been flown to Panama. Others have been flown to Puerto Rico, Cuba, and other Caribbean countries.

Deliveries to the West Coast are usually made by freight cars, since six disassembled ships can be loaded in one car, effecting considerable savings in delivery expense. Shipments to South America, Africa, etc., are by steamer and the planes are completely crated in a compact box. Cubs have been shipped to more than 60 different foreign countries. Before the war, Cubs were built at the Cub Aircraft Corporation plant at Copenhagen, Denmark, for distribution throughout Europe.

The December 1st Bulletin noted that, "Employee relations center around the Cub Fliers, an organization of workers interested in learning to fly. Organized in 1935, the Cub Fliers has grown to the point where it is now (in 1941) the largest flying club in the United States and boasts more than 300 active pilot-members (out of 1200 employees). It has currently seven airplanes, supplied by the company, and three instructors!"

During this period and later, Harvey Martin, aircraft salesman in Santa Ana, CA, solved the ferry problem by adding a large deisel fuel tank to his old Buick. When the engine was sufficiently hot, it would run on cheap fuel oil. Then he'd send three young men back to the factory, either Piper, Aeronca, Taylorcraft or Porterfield, and have them drive straight through. The makeshift trailer on the back of the Buick would handle two trainers with their wings removed. Fuel cost was at a minimum, there were no motel fees, and it's a good bet that the cross-country drivers were paid in minimal amounts of flying time. That's the way it worked in those days.

One of the senior birdmen in this general aviation business is Jim Dewey of Santa Paula, CA. His background includes running a flying school at the old Alhambra Airport east of Los Angeles long before WW II, handling a Civilian Pilot Training Program at the time of Pearl Harbor, flying C-46 transports over the Burma Hump, a long tour as a General Aviation District Office Inspector with the FAA and, more recently, enjoying the fruits of "retirement" by building and re-building classic airplanes (Fig. 5-15).

When we asked Jim about ferrying in the old days he blew some dust off the records and came up with this brief account.

"Of all the ferrying of antiques that has been my privilege in the past 50 years, the best was a trip from Alhambra airport to Santa Barbara Airport in 1932 or '33. I had just learned to fly with doubtful confidence, having enough experience in one of the few licensed Cycloplanes (C-1's) ever made to prompt the owner to let me deliver it to a new owner waiting at the Santa Barbara Airport at Goleta. (Same airport as now). The airplane has a 22-hp Cyclomotor which was two-cycle and two-stroke. Since it had an empty weight of 430 pounds and a huge wing (40 feet), the landing speed was around 20 mph in no wind.

My first cross-country briefing was conducted by the man who would profit by my safe delivery of the airplane. The briefing was verbal in its entirety. 'Fly west on Valley Blvd. until you get to the Los Angeles River—turn right and fly on the railroad tracks until they have a Y; be sure you take the left-hand tracks and stay on them and watch for them to go into a tunnel—fly over the hill and stay on the tracks until you see two round mountains—go right between them and you'll see Santa Paula pretty soon to your left. Land and get gas and don't forget to put the gas in first, then the oil; shake the wing real good before you start 'er up.

Fig. 5-15. Veteran pilot Jim Dewey at the controls of a gleaming Ryan STA that he rebuilt. Photos of the early Cycloplane are long since history.

When you leave Santa Paula, fly down the river until you get to the ocean and follow the shore to Santa Barbara and around the point to Goleta—and there she is."

Everything went well and I found Santa Paula, but no one was there. The runway was all covered with grass and weeds although there were quite a few hangars. I walked up to town a few blocks and bought five gallons of car gas and two quarts of oil which I poured in the tank while standing on a handy orange crate. I sloshed it all as instructed and soon was on my way, looking for the ocean. So, around the corner at Ventura and on to Santa Barbara, not needing to worry about fuel anymore since the Cycloplane burned about two gallons an hour from the two 3½ gallon tanks.

After what seemed like an endless trip beating upwind, I finally completed the assignment in three hours and forty-five minutes, which was not the speed the advertisement said. Anyway, the 140 miles in 3:45 was good enough. A freind in a Monoprep took me home. I couldn't hear for several days because the cylinders were just above my head, and you know what a two-stroke sounds like. I was too dumb to use ear plugs. Nobody did in those days 'cause that was sissy."

"Ferrying today is certainly a better proposition than it was at the end of WW II," commented Phil Oiler, a Tool Engineer from the City of Industry in California. He has owned the same 1947 Stinson Station Wagon, "Old Sturdy," for at least 14 years and advises that "we get along fine."

"A guy wanted to buy a couple of Fairchild PT-19's and have them flown to the old Monrovia Airport for rebuild, explained

Oiler. I was out of a job at the time and had put some Rawdon hatches on these airplanes, so I'd have a job with two planes to work on once the ferry flight was completed. Another pilot and I caught a car ride straight through with a fellow who was driving past Ranger, TX. There must have been 40 to 50 PT-19's on the field.

We picked up the two that the airport had purchased for about $500 each, checked them out casually and started west with an auto club map in each cockpit. I remember that it had been raining heavily in Texas and there was water dripping out the trailing edge of the wings. We borrowed some tape to patch the loose fabric before takeoff. When that tape was pulled off at the end of the trip, most of the trailing edge of the wing ribs came with it. The planes were a mess inside.

At Ranger, the FBO pointed out our ships and said cheerfully, "They're all ready to go." We were so hungry to fly that we didn't ask any questions. Surprisingly, both planes started right up and we headed back (Fig. 5-16).

Nobody was paid for that trip. We had fuel and meals covered and that was about it. Of course, I'd have a job for a few weeks rebuilding the two planes once they were returned. I can remember putting my top coat in the little baggage compartment because I expected the weather to get better as we came west. I damn near froze to death."

Oiler still remembers the trip because of the extreme cold. On his last leg of the trip into Monrovia, he stopped at Cable Airport, just 15 miles from his destination, to run in and stand over a stove for a few minutes to thaw out.

"In those days," reminisced Oiler, "we'd stop and ask the (then) CAA station about the weather ahead. It really didn't make much difference because we were going on anyhow. If it came down too far, we could always land on a back road. That was no big deal in those days."

Anyone for an open cockpit cross-country these days in the winter months?

How the Cubs Came West

How did you become a ferry pilot way back when? There was WW II and the U.S. Army Ferry Command, as noted previously. After that, many of us returned to the security of non-flying jobs while retaining an avid interest in airplanes. So let's go back to 1946 when Piper cranked out 6320 J-3 Cubs in that single year.

Fig. 5-16. Fairchild PT-19 takes off from a dirt road in Southern California after stopping for fuel. In those pre-marijuana days, few people bothered about off-airport landings. This PT-19 is similar to the model ferried by Phil Oiler.

Here's the story of one of those Cubs—NC-70171—and two ferry flights that bridged more than two decades.

I had settled back into the routine as newspaper photographer for the Pasadena, CA, *Star News* and was earning my flying time by moonlighting as Western editor for the old "Skyways" magazine. Old clippings and carbon copies show that the saga of 70171 began in this way:

The phone rang at about 11 a.m. and Jim Most, nearby FBO and Piper salesman, asked, "Can you take your vacation starting right now?"

"I can try. What's up?" I asked, already anticipating some flying.

"There's a car going east at 2 p.m. today with two pilots to pick up J-3 Cubs at the factory. I need a third ship brought out. Can you make it and take Ruthie with you?" Then he added, almost as an afterthought, "One of the pilots has 100 hours total time and the other about 200 hours. Maybe you could help a little with the flight planning and navigating." At the time, I had logged about 3200 hours—the bulk of it in Army aircraft.

There's a certain nostalgia in the respective costs of gas, cabs, meals and motels. Even then, the original $995 fly-away price on the Cub had soared to $2010! And just in case you didn't notice, the old photographs show considerably leaner people. Our total fuel cost on the 40-hour, five-day trip was $41.35 (Fig. 5-17).

There's always safety in numbers—and this Cub caravan sounded like an interesting junket. Fortunately, my city editor on the Pasadena *Star News* was a tolerant guy (for a city editor) and okayed the time off. Ruthie packed a single, small suitcase that we lived out of for the next nine days. I collected my aging Military Airways maps and therefore made my first mistake of the trip, before we ever left the house. Since there wasn't room for my 4X5 press camera, I borrowed a 2¼ × 2¼ folding camera.

The less said about the straight-through auto trip east, the better. However, we were lost on the ground more frequently than in the air. One rural gas station within 30 miles of Lock Haven had no one around who had ever even heard of the place.

However, after a long night's sleep in the corner room of the old Hotel Fallon, we walked into the lobby at Piper, presented our fly-away papers and were given the usual first-timer's tour of the plant (Fig. 5-18).

At this time, Piper was in full production of J-3's, and the lovable little yellow monsters were rolling off the assembly line at an all-time record pace. That's when we first met NC-70171 (Fig. 5-19). In those days, license numbers carried a "C" for commercial.

Pilots of the other two J-3's on our westward caravan were Hank Langley of Puente and Bill Baker of Riverside. Since there would be two of us in NC-70171, it was decided that we'd do the navigating. The other two planes would fly a trail formation,

Fig. 5-17. Back in the days immediately following WW II, you could fly a 65-hp Cub from the factory in Pennsylvania to Southern California for $41 in fuel. Remember when?

Fig. 5-18. Factory production at Lock Haven, PA, of the ubiquitous yellow Cubs following WW II.

somewhat lower and to one side so that they could keep us in sight.

With no radio equipment, we established that rocking the wings would indicate "yes" or "I understand." Yawing with the rudder meant "no." For anything else, we would fly a close formation and use hand signals.

The log books on NC-70171 showed a single 20-minute test-hop and the plane had that wonderful new-car smell as we stowed our suitcase, pulled out a set of maps covering Pennsylvania and a few points west, had one of the line boys give us a "prop" and "S'd"

Fig. 5-19. Parked J-3's awaiting ferry pilots at Lock Haven. Paved parking ramps were a rarity in those days.

our way to the head of the runway. Behind us, pilots Langley and Baker followed "mother hen."

We were able to make an early-afternoon takeoff (Fig. 5-20). A generous tail wind boosted us along at 100 miles-per-hour ground speed down the Susquehanna River Valley. Since the 12-gallon gas tank of the trainer isn't the best thing in the world for cross-country work—as we found out later in Texas—we landed for a refill at Pittsburgh's Pitcairn Airport after an hour and 45 minutes in the air. This grass field was rough, the latrine was drafty and dirty, and there was no place to eat, so we filled our tanks in a hurry and headed west. Since none of the trainers had even a radio receiver, we tried to stop at the smaller, outlying fields.

Following another fuel stop at Newark, OH, we pulled up for the night at Richmond, IN, as the car headlights from U.S. Highway 66 began to shine brightly in the dusk. Richmond was typical of many mid-continent fields: five miles and $1.50 taxi fare from town, no eating or sleeping accommodations on the field and insufficient hangar space. But it did have a CAA weather station on the field.

When we awoke the next morning, we couldn't see the 300-foot radio tower atop our hotel, so we ate a leisurely breakfast and sauntered out to the field when the weatherman dubiously called a 1000-foot ceiling and a scant three miles visibility through a powdery rain. Indianapolis, Terra Haute, Effingham and St. Louis rolled slowly beneath the wings as we bucked a stiff, 20 mile-per-hour head wind.

St. Louis on a Saturday night! We checked the hotels by phone from the Parks Air College hangar, just across the Missis-

Fig. 5-20. Two other J-3 Cubs follow the caravan west. Wide-open country and slow landing speeds make this sort of a trip relatively secure.

Fig. 5-21. Cow-pasture landings for fuel were almost routine with the 12-gallon capacity of the J-3 Cub. This pasture is near Jay, OK.

sippi River east of town. There were no rooms. We had a scant hour of sunlight remaining, and the map showed three fields grouped 25 miles west of the city. We crossed town, flying into a blood-red sunset. Transcontinental Highway 40 borders the Meramac Airport—and a neon sign spelled "cabins" not 100 yards from the hangar. As we circled, we could read "eat" on an adjoining sign, so we promptly landed. The steaks were a credit to the cattle country and the overnight accommodations quite adequate. This was the only field on the whole trip where we found overnight facilities within walking distance. Most airports had plans for "air motels," but the shortage of building materials had prevented their completion.

Navigation across Missouri was a cinch. The double band of Highway 40 stands out clearly and the other Cubs frequently broke away from our loose formation for a closer look at the farmhouses. We landed for fuel at Rolla and Springfield, then detoured through the beautiful Cherokee Lake section of northeastern Oklahoma, many miles from the main highway.

The hop from Springfield, MO, to Claremore, OK, was a bit long for the winds we encountered, so we all landed in a cow pasture near the little town of Jay, OK (Fig. 5-21). The Cub's slow landing and quick takeoffs are really appreciated in flying from unimproved fields. We each purchased five gallons of auto gas from the local station, dumped it aboard and continued to Claremore.

There was another overnight stay in Shawnee, OK, where our two accompanying pilots dropped off to visit friends. We continued westward alone—and a little lonesome.

Duncan, OK, Wichita Falls, Seymore and Sweetwater, TX, passed before complete darkness caught us again. Avenger Field, former home of the WASPS (Women's Auxiliary Ferrying Squadron—W.A.F.S.) was a spacious airport with plenty of hangars and the best of service. The day after we arrived home, a postcard came from the manager inviting us to stop again on our next trip through Texas.

Once more we planned an early takeoff and were delayed by rain squalls and high winds. The ceiling lifted to a bare 1000 feet, and we gingerly wheeled the little Cub out into a gusty 30-mph wind. If it had been a head wind, the Cub would have stayed on the ground; but 30-mph tail winds from east to west are a novelty in Texas and too good to miss. Almost as soon as the mechanics let go of the struts, we were in the air. Our ground speed to Odessa was slightly over 100 mph.

The winds calmed as the day progressed, and we followed the airways toward El Paso. Our aging ex-military charts showed a thriving airport at Wink, TX. When we arrived, the airport was completely closed and no fuel was available. However, we still had considerable fuel, so we headed down the diminutive pipeline toward El Paso.

As the gas gauge dropped lower and lower, we finally passed over an isolated filling-station, post office and general store. There were no telephone wires adjoining the road, and only an occasional car, so we circled once and landed on the paved highway. The "town" was Orla, TX, midway between Pecos and Carlsbad Caverns (Fig. 5-22).

We taxied right up to the gas pumps and filled up with auto gas. The little Continental engine ran as well on this diet as with 73-octane aero fuel. The proprietor of Orla seemed not at all impressed with an airplane at his gas station and commented, "Shucks, you're the third plane that's been here this week."

We took off from the narrow highway and headed for El Paso. The air is normally quite turbulent over the desert in summer months and this day was no exception. On one occasion, our lone suitcase and all loose maps floated to the top of the cabin as we hit a patch of rough air near Guadalupe Peak, the 8750-foot landmark east of El Paso.

With a 12-gallon gas tank, you don't pass any fueling stops; so we landed at the CAA emergency field at Salt Flat, topped-off again and flew another 55 minutes into El Paso. One thing that you can say about a J-3 delivery that doesn't apply these days; with the

Fig. 5-22. Orla, TX, was a popular highway stop for Cub pilots. This J-3 was the fourth airplane to fuel there that week. An up-to-date caution: Modern aircraft engines are NOT designed for auto gas, although the old 65-hp Continental did very well on this liquid diet.

J-3 you had plenty of takeoff and landing practice. We logged 34 landings on this 5-day flight.

The municipal field (now International) at El Paso (Fig. 5-23) had a good lunch room. After filling our gas tank and ourselves, we

Fig. 5-23. J-3 Cub at El Paso International Airport beside one of the DC-3's used by the airlines in those days. No radio was required as long as a flight plan was on file. Don't try this procedure today!

returned to the J-3 to find a note from the control tower requesting that we please make a left-hand pattern around the field. We'd followed an American Airlines DC-3 into the field on a right-hand pattern. However, the tower gave us a green light and we shook the dust of Texas from our landing gear without further trouble.

We followed the railroad tracks, just three miles north of the International border toward Columbus, NM. As far as the eye could see, there was nothing but sand, sagebrush and an occasional red-rock volcano core.

Flight time to Cilus Was only 55 minutes, but four gallons of gas was important in this part of the country. We paid an all-time high—35 cents per gallon. At our request, the CAA weather forecaster teletyped ahead to Rodeo Radio, an hour and 15 minutes by Cub, to make sure there was fuel on the field (Fig. 5-24).

After fueling at the isolated Rodeo Radio emergency field (Fig. 5-25), we headed into the setting sun for Tucson, AZ. Our maps showed an airport at Benson, AZ, but when we arrived over town the field was grown over. Decision: to go into Tucson after dark on a black night and without lights, or find a friendly field near Benson (Fig. 5-26). We circled town, picked out a broad ramp in front of three adjoining service stations and "planted" the Cub for the night. Gas, a cafe and motel were all nearby. The CAA normally frowns on highway landings, but in the Far West where gas stops are scarce and there are normally no wires, they classed these landings as "an emergency." And when that gas gauge is bouncing on the bottom, it is!

Fig. 5-24. WW II B-29 is parked with an engine "out" at the dirt emergency field at Columbus, NM, as the J-3 Cub stops again for fuel.

Fig. 5-25. Rodeo Radio, an isolated CAA radio station, had a few cans of gas available. With the advent of VOR's, all these intermediate stations have been closed.

Don't try a highway landing today unless that powerplant has quit cold and you can prove it—or you're way out in the back country. The FAA, as well as State and County law enforcement officers, have regulations that make highway landings illegal. If such a complaint is filed against a pilot by either state, county or city officials, the FAA is forced to file a violation. Yet many FAA

Fig. 5-26. Once upon a time there was no airport at Benson, AZ, and several pilots landed on the main highway, taxied up for auto gas and continued. **A repeated reminder: Do not use auto gas in modern aircraft.**

flight examiners will say, "If I'm in a real emergency, I'll put down on a road first and argue about it later.

The owner of one of the Benson gas stations was a former Army Liaison Pilot. He gave us "a prop," drove ahead to block off traffic, and we were on our way. Years later, we stopped our car for fuel at this same gas station, and the operator still remembered the J-3 Cub that had spent the night there.

Next to our last day, we landed at Casa Grande for gas and then stopped at the huge AAF base at Dateland. It was the same old story. The field was abandoned, except for a watchman, and there was no fuel. However, a small gas station and store at Tacna had a fairly smooth field nearby—there were wires along the highway—so we filled up and headed up the Colorado River toward Desert Center. West, on the horizon, were the towering peaks of San Gorgonio and San Jacinto. We were nearly home.

The next day, NC-70171 was delivered to the Montebello Airport where she piled on over 5000 hours as a trainer before Jim Most sold here in 1953. FAA records show that two owners and two years later, William J. Campbell purchased the ship and, in October, 1956, converted it into a fish-spotting plane. Mechanic Robert J. Tobert of the Orange County Airport, Santa Ana, supervised the modification.

The History of NC-70171

It's always interesting to back-track the goings and comings of an old friend like N-70171. "I recall my first contact with 'your Cub'," said Mr. Tobert. "I was called to a house to inspect a Cub , prior to a do-it-yourself recover job performed *a' la* garage. The owner (William J. Campbell) was an ex-commercial fisherman who was building up the Cub to work as a fish-spotter with the fishing boats. This Cub was scattered all over his garage in various stages of dismemberment. One wing was strapped to the overhead rafters, the other wing was on saw horses; tail surfaces were nailed flat to the walls; the fuselage—sans landing gear—was in the driveway where it had been put for my inspection. I later inspected the airplane again for license and signed off the inspection of fish-spotting gear in it. This included an Apelco marine-band radio, trailing antenna and reel, a 26-gallon Ryan PT-22 gas tank mounted on the rear seat for extra fuel and a small 6-gallon wing tank for more fuel. Of course, there was a wind-driven generator and battery installed to provide power for the radio and fuel transfer pump.

"William Campbell later decided to sell out and removed his extra gear to make the plane more salable," continued Robert Tobert. "I, of course, had to sign it back into the Standard Category. The airplane did not sell for a long time—just sat there going to pieces. Finally, a mechanic who worked for me, Donald Pridham, purchased the plane with another fellow, and they flew it some after more-or-less overhauling it, including, I believe, an engine change. Rust had ruined the other engine. Don finally sold the Cub after recognizing that he could not support a growing family and an airplane, too. My knowledge of this plane ends at this point."

The ship then went to Richard Weis of Pomona and three co-owners. Late in 1964, it was purchased by its present owner, Arthur Chap, Jr., for $1115, over 6000 hours and nearly 20 years after it had been sold new for $2010. Mr. Chap also has a Cessna Skylane, but he purchased the old J-3 "just as a fun airplane." He also worked on his Instructor's rating, using the J-3 to brush up on techniques and maneuvers so that he could *really* teach some of his friends how to fly (Fig. 5-27).

In browsing through the photostats of the FAA records, an interesting inconsistency came to light. An old CAA annual aircraft inspection report dated September 21, 1952, stated: "Records indicate that the airframe has been operated a total of 6225:45 hours." A similar CAA form dated over two years later on September 29, 1954, showed a total airframe time of only 5449:55 hours. That's a nice trick, if you can do it.

In late 1963, the FAA inspection report showed a total of 6361.8 hours. However, what's 800 or 1000 hours more or less for a rugged old bird like N-70171?

However, we aren't finished with N-70171. It is seldom that a pilot has the opportunity to ferry the same airplane twice. It was 20

Fig. 5-27. The same J-3 Cub shows up years later and is flown to the Palm Springs, CA, airport on a visit.

years later that I made a second long delivery trip with this J-3. During the intervening years, she had been in Southern California. Arthur Chap, Jr., was kind enough to let us fly it a couple of times, and I gave daughter Dana some of her first dual instruction in the ship.

Return of an Old Friend

Our second ferry flight in this J-3 Cub began like the first—with a phone call—this time from Pat Chap. It seems that he'd sold the ship to his brother-in-law in Chicago and wasn't too enthusiastic about taking the time to deliver it. Dana needed the flight training, so I arranged to make a return ferry flight with a new Cessna 150 which kept our expenses to a minimum. Chap's brother-in-law agreed to meet us in Wichita.

Subsequently, Dana's impressions of the ferry flight saw print in *Air BP*, an international aviation magazine out of London.

She wrote as follows:

I must have been an idiot going cross-country in that airborne piece of junk! My Dad referred to it as, "the loveable yellow monster," but it wasn't yellow any more—sort of mustard color with a dash of orange, and it wasn't really lovable (See Fig. 5-28).

I had my Student Pilot's permit before we started out on the trip and Dad had me flying in the front seat all the way. But there's a lot more to flying one of these old taildraggers than going around the pattern in a new tri-geared trainer.

After we had packed our baggage aboard, I climbed up front while Dad "propped" the old wooden propeller. Of course, the first time I didn't have the mags on and was promptly told about it. Off one wingtip, Mom laughed to herself (Fig. 5-29).

Finally we got the thing started, waved goodbye and taxied out for takeoff. There was no radio, but we'd called ahead to the control tower at Brackett Field, and when they discovered it was a delivery flight out of the state they were most co-operative with the red and green lights.

I managed to stay inside the runway on takeoff but 171 didn't climb like any ball-of-fire. Those tired 65 horses, sticking their sparkplugs out the sides of the cowling, had seen better days and we didn't have enough altitude to get through Cajon Pass until we were in the middle of it.

It's normally a quick 190-mile hop from our side of Los Angeles to Las Vegas, NV, but we stopped halfway at Daggett for fuel. That was just the beginning. The next day shortly after noon, we took

Fig. 5-28. N70171 heads east again after more than twenty years.

off—just barely—from the airport at Holbrook, AZ. The day was hot, the altitude 5245 feet, and we were five miles off the end of the airport, flirting with the sagebrush, before we were high enough to make a turn. That 65-hp engine was really tired (Fig. 5-30).

However, Dad is an old-time sailplane pilot, and he spotted the beginning of the first cumulus cloud of the afternoon. We flew away from the main highway, where the trucks had been passing us, and headed in under the cloud. Dad made a couple of turns to get the thermal centered and up we went. Then I took over and tried to keep circling in the rising air. What do you know—soon we had

Fig. 5-29. There's just no easy way to get into a J-3 Cub.

10,000 feet and headed eastward toward the Continental Divide. It took a number of thermals to get us over the mountains before we landed for the night at Grants, NM.

Grants is high, 6527 feet; so we left a call for well before dawn and took off while it was still cold. It had something to do with the density altitude.

Even with the six-gallon extra wing tank in operation, we didn't have enough fuel to get to Tucumcari, so we landed at Albuquerque and topped off with just four gallons. Then we started to climb—slowly, of course, and flew all the way north of the Sandia Peaks to get enough ridge wind to hold an altitude of 8000 feet. We had a good look at the cows until the country started going downhill.

After that, we had flat country all the way and the perspiration was gone. Dad asked where I'd like to spend the night, so I spotted Dodge City on our Sectional charts. If you've ever watched some of the old cowboy movies on late-night TV, you've heard all about Dodge City—so that's where we headed.

We could see storm clouds chasing us from the west, so we had the old Cub put in a hangar that night. Just the slightest bit of hail would have really made rags of that rag-wing.

There was only one fuel stop the next morning before we landed at Wichita's Piper Air Park. As we landed, Arthur Chap was circling the pattern, coming in from Chicago with his brother-in-law and he taxied right up beside us in his Cessna Skylane (Fig. 5-31).

This had been a real for-fun trip for me. People had come out of airport offices all the way across the country to look at the old Cub and ask how we were getting along. When we flew a brand-new Cessna 150 back home, no one paid the slightest attention.

The Tale of this Cub Has No Ending

Actually, when Edward Potter took off for Chicago, it was the last time we saw N70171—but the story doesn't end there. Mr. Potter had only 12½ hours total lightplane flying, plus some 200 hours as an Aviation Cadet 22 years earlier. He viewed the impending 600-mile cross-country trip in the J-3 with mixed emotions, particularly after he learned that the compass was wildly inaccurate, that there was no radio, no starter, no lights nor gyro instruments and a minimum fuel capacity.

He later wrote and told me how, "Art and I shot landings for about an hour and he finally 'allowed-as-how' I might be able to get it off and back on the ground again without wrecking it. Apparently having less than complete faith in my navigational ability, Art re-

Fig. 5-30. Dana in flight with the J-3 Cub. Note missing instrument holes in the instrument panel.

commended that I follow the Kansas Turnpike and other obvious and well-marked highways back to Chicago.

"The next morning at 8:10, I soloed the old J-bird and kept right on going. Slowly, very slowly, as befits a lady of so many

Fig. 5-31. The J-3 gets a final "prop" before delivery in Wichita. Pilots changed seats on each leg of this delivery to stretch.

summers, she climbed up to 3500 feet, which I had selected as my cruising altitude.

"After about an hour of uneventful flying I began to realize how rapidly that piece of bent wire was sinking into the gas tank, and an hour later I 'chickened out' when only three inches remained and turned on the six-gallon auxiliary tank. This proved to be quite a gymnastic feat, since the selector switch was located under the left side of the instrument panel and I was flying from the back seat!"

Edward Potter completed the 598-mile flight in a flying time of six hours, 29 minutes, and five weeks later wrote again to relate how, "Fate wrote another chapter for old 171. A violent storm struck last Thursday night, with winds at the airports up to 90 mph. The old girl was chained and didn't go anywhere—some planes flipped over on their backs—but the downward pressure of the gusts was too much for her poor old wings and the left one collapsed. Now she looks like a deflated barrage balloon. The insurance is still in effect, but it appears that it will cost more to rebuild the old girl than she's worth."

That should have ended the story, but 171 is a dogged old girl. Her damaged airframe was bought by Nick's Aero Service for salvage. One of Nick's part-time employees, a United Air Lines 727 crew chief-mechanic named Bill Flynn, takes up the story: "I had seen a Cub perform at an air show last year and since then had been looking for a J-3. I helped rebuild 171 and when it was finished I bought it from Nick for $1100.

"My flying experience is nil. I'm just barely approaching 200 hours and the Cub just filled the bill for getting 100 of those hours. I'm using the Cub to build time and put in a dozen hours a week on it. I'm also rebuilding a Tri-Pacer and plan to sell the Cub for overhaul money for the Pacer's 150-hp Lycoming."

As the Tri-Pacer neared completion, 171 was sold to Carson E. Thompson of Elmhurst, IL, who has told us that he learned to fly in a J-3 and this started my fondness for the aircraft. "I bought the Cub just to have one to fly and the only thing I've done to it was to replace the rear seat with a standard J-3 seat. This fits two youngsters very nicely and we've had a ball bouncing around the area.

"As for the future, I just plan to fly the aeroplane whenever I can and keep it 'till I get a homebuilt Starduster flying. Then I suppose I'll sell the Cub, but who knows!"

And after all these years, I wonder if that particular old J-3 Cub is still putt-putting around the skies?

Chapter 6

Used and Abused

Some airplanes age well with "TLC" and good maintenance. Others that you may ferry have been sitting for months or years. Approach them with a great deal of suspicion. It's only good insurance to review dusty operations manuals and talk with pilots who are familiar with that particular brand of beast. Modifications and restorations have their own peculiar problems.

Bitten by a Canadian Beaver

Jack Leggatt, who recounts this ferry flight with a surplus Canadian Beaver, is a Canadian by birth and a pilot since 1946. He has spent the past dozen years in Southern California where he edited *Plane & Pilot*, *Sport Flying* and other popular general aviation publications. It has been the author's fortune, and occasionally, misfortune, to share the same cockpit with Leggatt on several occasions including the delivery of a much-modified Cessna 175 CF MFK from Southern California to Edmonton, Canada. When we ran off the end of Leggatt's Canadian charts, the arrow on the highway pointed toward Edmonton—90 miles. Leggatt laughed and said, "Just stay with the highway. We'll get there in time for lunch."

Nothing sounds more glamorous than a ferry flight. Lots of fun flying, all expenses paid and a trustworthy new airplane to fly. Not all flights are like this—in fact, very few ferry flights are.

How about grabbing a well-used, and tired, military surplus airplane from a military base and heading some 2000 miles cross-country? The airplane had been sitting for four-plus years, and all that was done to it was a "quick" 100-hour check and engine run up.

Being a coward, I have never crossed the Atlantic or Pacific in anything less than a Boeing 747, cocktails and hostesses included. But I have stuck my neck out—once—and ferried a military surplus de Havviland Beaver six/seven-seater bush airplane from Tucson to its new home base in British Columbia, Canada. The experience was enough to make a Christian out of me, if I wasn't before that hair-raising flight a couple of years ago (Fig. 6-1).

The problem started when a used-airplane salesman from Edmonton, Alberta, Canada, Mick Hackman, talked me into ferrying one of his six newly-auctioned-off military surplus Beavers. "No sweat," he told me. "We'll give them a fresh inspection prior to your takeoff and you will be on your way."

Little did he, or I, realize the hidden complications of the forthcoming trip. Looking into my diary, I recall some of the highlights, or "lowlights," as I sometimes call them. To this day, I still use a momento of the flight—a pair of engine-oil-soaked leather shoes—when I work in the yard or under a grease-covered automobile.

The first leg of the ferry flight started across the fence surrounding the Davis-Monthan Air Force Base just outside the Tucson city limits (Fig. 6-2). It ended 10 minutes after lift-off, and I was still in the city limits—and oil-soaked to boot.

Please allow me to go back to my first look at the Canadian-registered de Havilland Beaver, C-GJAK. Canadian registration is

Fig. 6-1. Moth-balled deHavilland Beavers sit near Tucson, AZ, before auction and subsequent ferry flights. That's not snow in the windows; it is preservative, now hardened so that it takes two days of careful peeling to remove it. A spray bomb of black paint has removed the military insignia. The air scoop on top of the cowling is for desert operations (courtesy Jack Leggatt, Figures 6-1 through 6-6).

Fig. 6-2. A much-weathered deHavilland Beaver in the desert near Tucson, AZ. Note preservative on the windows.

either CF-AAA or C-AAAA, and upwards. It was sitting in the main area of an FBO who specialized in overhauling and making flyable military surplus airplanes that have been purchased from the U.S. Government and stored at the massive Davis-Monthan Air Force Base (Fig. 6-3).

The FBO's runway is a little-used road, immediately adjacent to the storage yard. After a plane is auctioned off, it is usually towed off the Air Force Base, and here the art of making a moth-balled airplane flyable again begins. The project can take as little as a week, or months, depending upon the condition of the engine, airframe and FAA paperwork.

C-GJAK was tired; at least that was my first opinion. I was later proven to be right. The paint was shot. The Army-drab paint job was flaking off, faded in many areas. Years of sitting in the hot Arizona sun had baked the moth-balling window protection cover-

Fig. 6-3. C-CJAK after $10,000 worth of modification and repainting.

ing into an almost impossible film that had to be removed before flight time.

Hackman and his partner and fellow ferry pilot, Ken McHeffey, had been working for two days in removing the window protective covering (Fig. 6-4). Meanwhile the local mechanic had drained all the engine inhibiting oil, installed new oil, and was in the process of installing 18 new spark plugs and a new battery. Then the stagnant 80-octane fuel had to be removed from the five tanks—two wing-tip and three belly fuselage tanks.

It was after fuel draining and system flushing that the first problem developed. Fuel was running out the tail wheel area. Something was amiss; there is no fuel dumping valve on a de Havilland Beaver.

The trouble was finally traced, three hours later, to a deteriorated fuel hose connecting the belly tanks. Four years of heat had taken its toll on this item also.

A normal inspection of the hydraulic system, cables, etc., showed a little wear from the last inspection. Theoretically, everything should have been in working order when the last Air Force Sergeant signed the airplane off to the storage depot inventory. Theoretically is a pretty good word—it could cover a multitude of sins. In C-GJAK's case, it did!

Finally, it came to run-up time. The enterprising mechanic fired up the R985 Pratt & Whitney Wasp, Jr. engine and with a belch of smoke and several nerve-shattering backfires it settled into the normal P & W noise pattern.

For 10 minutes, the engine was warmed up. Then it was shut down and an all-important oil change was made. All the inhibiting oil must be flushed away.

After watching all this activity, I was starting to feel that everyone was doing a good job and that my departure the next day was to go as scheduled.

But before I left that final evening I sat in the cockpit. "My gawd!" I said to myself. Fifteen to 18 hours in that military-designed seat that would fit a 150-pound , second lieutenant pilot, complete with seat parachute, would give me "piles" of problems.

A stop at the local furniture store would produce three inches of foam rubber for satisfying comfort, I thought. That proved another gross error later.

Bright and early the next morning, padding in place, I fired up the R985 and carefully checked all pressures, temperatures, etc. It was a very meticulous ground check, so I thought. Feeling cocky

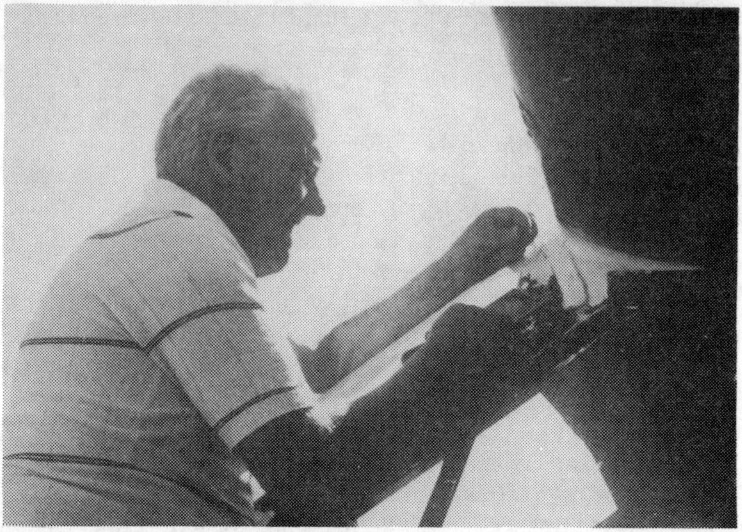

Fig. 6-4. Ferry pilot Ken McHeffey, painstakingly removes window preservative that is case-hardened to the plexiglas after years of desert heat.

after another 15-minute check of mags, hydraulic systems, etc., I accepted the bird as airworthy and made immediate plans to head northwest to Las Vegas, my first refueling and rest pause.

I had thought of a test flight, but was advised not to for several reasons. The prime one being that the FBO's airstrip was only "unofficially" recognized by the nearby Tucson International Airport controllers and the tower brass at the two-mile-away Davis-Monthan Air Force Base.

"We suggest you take off, make a slight turn to the left and pick up the main highway on your northbound route," said the knowledgeable mechanic. "The highway separates the two control zones and if you stay below 800 feet, there won't be any problems," he added. "They don't like somebody flying the pattern."

That made sense to me. After all, there is lots of desert around if I have to make an emergency landing. The Beaver is a STOL airplane. With just myself in it, it should land in less than 200 feet—that is, if you have a hundred hours in a Beaver. My total experience was less than five hours; but then, I knew everything!

So with a dozen Sectional maps in my hand, I climbed into Hackman's $24,000 military-surplus Beaver and headed north.

The takeoff was astonishing. Why I used full flaps (as the Book suggested) and full power, I will never know. As I shoved the wheel forward to raise the tail, the eager Beaver and I left the ground. Hackman later said I took 200 feet to get into the air.

Throttling back and pumping up the flaps (yes, the flaps have to be pumped up), checking rpm and sliding over to the freeway, I was somewhat busy. A couple of the latest jets skyrocketted off the one-and-a-half-mile-away air force base for a sortie. I dropped back to 800 feet per instructions. This was an ex-Army airplane, and I didn't want to tangle with any Air Force students out to prove their manliness.

Downtown Tucson was just ahead and things were warming up—in more ways than one. I was warming up to flying this beast, and it was warming me up.

The cockpit heater valve was closed, so it was open-the-sliding-window-time to cool off. It still got warmer and warmer. Oil temperature appeared well below the red mark and cylinder head temperatures looked reasonable.

But something was amiss. It was getting hotter in the cockpit. My feet were getting warm, also. About this moment, my foot slipped off the rudder pedal into the military-designed heel well. I then glanced down and got the shock of my life.

The heel wells were full of oil. My shoes were covered with blood-temperature oil. Oil was running down the cabin floor. At the rate it was flowing through the cockpit, the five-gallon oil tank was good for about another three minutes.

A frantic check of the map showed Freeway Airport, just at the north edge of Tucson, was about two minutes away. It was the shortest, straightest final that that Beaver had ever been on. The Aztec pilot turning onto takeoff position must have decided it was another bad day. The Beaver went over him at 15 feet and it was a panic landing, to say the least.

As I hit the ground on one of my numerous bounces, the emergency guel and oil cut-off valve was pulled and my flying day was over (Fig. 6-5).

It took Hackman and mechanic a couple of hours to round up transportation, repair equipment and get to the new "home base" for C-GJAK 10 miles away.

I really didn't care at this point what the problem was and I didn't care if they ever fixed it. But the next morning, I had a fresh outlook on life and the forthcoming ferry flight. I met the mechanics at the grounded Beaver.

"Must be a blown oil radiator," suggested the mechanic. He was right and he refilled almost five gallons of oil in the cockpit-located oil tank and started up the engine.

Fig. 6-5. All that black shiny stuff is oil, leaked after the Beaver came to rest at the Freeway Airport, Tucson, AZ. Even the ground was covered.

The problem was traced to the final inhibiting services performed by some enlisted man—possibly the sergeant who signed the plane off to storage. For some reason, he had disconnected the oil radiator, drained the oil, and then inserted a plastic plug into the oil line—no doubt to keep contaminants out of the engine system. The he reinstalled the lines.

Naturally, in a 15-minute runup, the oil didn't reach normal operating temperatures and the bypass valve had no reason to open. After 10 minutes of brief climb and cruise power, however, the temperature got to normal and the bypass valve opened—obviously to no avail because of the stopped oil flow. We came to the conclusion later that the plug got hot and melted, jammed the bypass valve and allowed pressure to build up in the oil radiator, which eventually blew (Fig. 6-6).

Whether this problem could have been foreseen prior to flight was a subject of discussion many times later. A simple "once-around-the-pattern" flight may or may not have proven anything; if it was a five-minute flight, probably the answer would be a "no." A fifteen-minute test flight, obviously "yes."

Fig. 6-6. While tow mechanics remove the malfunctioning oil radiator from the DCH-2 deHavilland Beaver at Tucson, Ken McHeffey, left, and co-owner Mike Hackman watch with interest. Note oil all over the side of the airplane and on the ground.

Needless to say, the rest of the flight to Canada was trouble-free—mechanically speaking. The seat proved to be a pile of problems, but a car cushion at Las Vegas solved that problem. The heater didn't work and the obsolete military radios never even produced static. Snow fell overnight in Reno and it took three hours to pre-flight that problem.

But that oil radiator! I still have the remains of that little plastic plug on my mantle. When I think I know it all I go over and pick it up (Fig. 6-7). Then, I'm a believer all over again.

More Fun than the Airlines on a Clear Day

Some ferry flights are taken on the spur of the moment. A quick run down the Pacific Coastline in a Cessna 150 serves as such an example.

I was sitting in Henry McKay's office at Robertson Aircraft, Renton Field Seattle, WA, one clear summer day. In my pocket was an aircoach ticket to return to Southern California late that evening after covering a series of airspace meetings in the area.

"The weather's good for the next day or so," smiled "Uncle" Henry McKay, vice president of sales for Robertson and a veteran of one bail-out from a Cessna 150 near Seattle. "How would you like to take a 150 back to the Los Angeles area and I'll have the owner fly in from Hermasillo, Mexico and pick it up?"

If you really like to fly, cross-country, even at the speeds of a trainer, is much more enjoyable than strapping on the coach section of an airliner and getting home in the middle of the night. McKay pulled a photocopy of my pilot's license and medical for his insurance form, handed me the keys to XB-DAS and pointed out the 150 on the ramp of the Renton Airport.

For some obscure reason, the Mexican owner had invested in a complete Robertson STOL kit and arranged to get the airplane to Seattle without first either majoring the engine or installing a larger powerplant. Therefore, I was flying a fine STOL airplane, but a ship without the takeoff and climb performance to really warrant the modification.

When the weather in the northwest is good, it's *great*, and this was one of those afternoons. So I borrowed a set of Sectional charts from McKay—Seattle, Klammath Falls, San Francisco and Los Angeles—and debated about his offer of a portable radio. It was a battery-powered unit, bulky and of a mongrel manufacture. When McKay promised that I could leave it with the airplane and he would arrange for its return, I took it along. It would have been less trouble to have left the unit in Seattle because it never worked.

I putt-putted out of Seattle, soaked in the vibrant beauty of Mt. Rainier and the subsequent snow-capped peaks and finally stopped at Albany, OR, for fuel and to check on any possible oil leaks. That 100-hp Continental up front had been exposed to a straight diet of Mexican small-town maintenance, but it ran reasonably well.

Fig. 6-7. Canadian Jack Leggatt will laugh at almost anything, even a boot full of hot engine oil.

Then I was on to Roseburg, OR, where I ran out of daylight. The motel here is close to the airport and they will pick you up—that's always a plus in an overnight stop.

Then it was on south with no blazing speed to Redding, the Nut Tree for a good and expensive lunch, San Jose, Santa Ynez and Chino near home. The trip had taken an extra day and I pocketed the value of my aircoach ticket in exchange for a one-day vacation.

A couple of days later, McKay called and asked if I could conveniently take the ship on to Phoenix or Tucson where the owner could pick it up. I called a former student, Bob Coffey, and we went on over to Arizona with a fuel stop at Blythe, CA.

Undoubtedly, the Robertson STOL modification proved to be useful in some of the smaller fields in Mexico. When the portable radio didn't work again, I left the ship at Deer Valley near Phoenix and put the radio and the keys in the Airport Manager's Office. Then Coffey and I caught an aircoach back to Los Angeles.

This was an informal, unplanned, kind of fun sort of a ferry flight that happens every so often.

Oldies but Goodies

Take two old airplanes—actually one was a restoration—and ferry them most of the way across the country, and you just can't help but come up with an interesting trip. This two-plane junket involved a sparkling two-place Fairchild 22 powered with a Menasco 125, 134-hp engine that was originally built in 1934. The second ship was a 1931 Buhl Bullpup that was originally powered with a three-cylinder Zekeley engine that had a propensity for throwing complete cylinders in flight. N320Y now has a 65-hp Continental up front.

Pilot age and experience were contrasting. In the back pit of the Fairchild was 72-year-old Jim Dewey, a 25,000-plus hours pilot, who had rebuilt the airplane from a nameplate and a handful of worn-out parts. Flying the Bullpup was John Peterson, age 22, with one year and 380 hours of flying time. Yet this was his sixth long delivery flight (Fig. 6-8).

Young John Peterson outlined the trip for Plane & Pilot magazine and gives his first person view of the trip as follows.

I'd worked on both of these airplanes and had flown them both before a customer, Bob Fergus, at the Union County Airport 25 miles northwest of Columbus, OH, purchased them. I'd been scheduled off and on to fly the 'Pup' if Mike Dewey couldn't take

Fig. 6-8. John Peterson warms up the Buhl Bullpup at Santa Paula prior to a cross-country delivery flight.

the time away from the office. Fortunately for me, things were busy and he stayed home.

Jim Dewey took the Fairchild that cruised up to 110 mph but throttled back to 90 and I was assigned the Bullpup that would go about 75-80 mph. I took a ride with Jim Dewey a couple of months before this ferry flight in that beautifully restored Fairchild. After a couple of landings at Santa Paula, he climbed out and waved me on my way. It's a great airplane to fly and that big, soft landing gear solved any touchdown problems.

We loaded our baggage and I made sure that my suitcase was tied so that it wouldn't slide back and cause an excessive aft center of gravity problem. We had already added a temporary six-gallon ferry tank located just back of the seat between two braces that hit me in the back. The Pup burned about 4½ gph and we figure 15 gallons usable, including the auxiliary tank. The main tank of the Pup has ten gallons and we carried six more in the temporary tank. This tank and the wobble pump were removed when we delivered the plane.

I'm 6'4" and don't fit well in some cockpits. However, the Buhl Pup had a huge cockpit where you sit on the bottom of the fuselage, and I wasn't cramped. It was a little cold on the first two hops but not at all bad at the time of the year (June) that we made the ferry flight.

Mr. Dewey did the actual flight planning (Fig. 6-9). He showed me how to put a heavy line on the Sectional charts and how to slide them under my leg so that they wouldn't blow out of the cockpit. The Pup didn't have a radio and the portable transceiver in the Fairchild was fresh out of batteries, so it was strictly a pilotage trip. We made no effort to stay in sight of each other since the Fairchild was about 30 mph faster than the Pup. Mr. Dewey was usually fueled and waiting for me at each stop.

We took off from Santa Paula, CA, and our friend Phil Oiler, who helped rebuild the Fairchild, flew escort as far as Fillmore with his Stinson (Fig. 6-10). Then he peeled off for home and both of us settled down for a long, slow look at the countryside. The first leg of the trip was bumpy, but our slow cruising speeds made the moderate turbulence seem pretty light. There were some high scattered clouds over the Santa Clara River Valley with visibilities down to about three miles, but that's par for the course anytime you get within sniffing distance of the Los Angeles Basin.

Visibility improved as we skirted the Edwards Air Force Base restricted area and headed for Daggett. Navigation here was no problem because of the broad Mirage Dry Lake where they fly gliders and the main Los Angeles-Las Vegas highway that comes through Cajon Pass and goes right near the Daggett Airport. However, there were dust storms blowing up from the dry lakes and I was a little apprehensive of high surface winds. Fortunately, the winds were light when I landed at Daggett with just two hours in the air from Santa Paula.

Mr. Dewey already had the Fairchild fueled and tied down. A nice lady from the Calico Motel picked us up and drove us back for

Fig. 6-9. John Peterson, 22, and Jim Dewey, 72, discuss the flight of two restored antiques at Santa Paula. Peterson flew the Buhl Bullpup and Dewey was in the faster Fairchild 22.

Fig. 6-10. John Peterson, in the Bullpup, takes off first from Santa Paula, followed by Jim Dewey in the Fairchild 22.

dinner and overnight. Mr. Dewey and I looked over the charts for the next day's flight and turned in early.

We were off the ground by 7:45 a.m. and headed over the hill toward Kingman, AZ. As the day became warmer, some high scattered clouds began to form. There's no tailwind that we had expected—perhaps even a slight headwind—but the Pup is running neat. There was a smell of gas fumes in the cockpit that began to annoy me so I flew with my head out the side of the glass.

Mr. Dewey is somewhere up ahead as we pass over the Colorado River—so long, California!—close to Needles. You can see the big red needles along the river (that's what they named the town after) and Lake Havasu farther down river.

I can see smoke over the hilltops and leave the desert highway for a straight shot at Kingman. The shortcut saves me 15 minutes, but it took me over some very hot desolate country. An engine failure here could produce a vacation on a mesatop with a steady diet of lizards and Gila Monsters while getting very thirsty. However, that's one of the saving graces of a two-plane ferry flight. If I didn't show up in Kingman, it wouldn't have taken Mr. Dewey very long to make a couple of phone calls and have someone searching the area for me.

I circle over Kingman and see the Fairchild at the gas pumps. I land and taxi up to the pumps, put chocks under the tires since the Pup doesn't have any parking brakes and we try to find something to eat and drink. Of all the times we have been in Kingman that restaurant has never been open. After a sip of water, we're back in the air and I see Jim disappear in the distance. Now it's just me and Mr. Buhl. Too bad we didn't have a tow rope to hang on the Fairchild and pick up my speed with the Pup.

Our next stop is planned for Winslow. As we fly along over highway I-40, the cumulus start to build up and it begins to get choppy. Some huge cumulus blossom over the San Francisco peaks just north of Flagstaff. Did you know that many Arizona-based pilots use those high peaks as their No. 1 navigational aid? It is seldom IFR in that part of the country and the 12,635-foot peaks seldom stick into the base of the clouds. This is one place where Dr. Paul MacCready—Gossamer Condor and Gossamer Albatross designer—did much of his early scientific research in the care and feeding and control of cumulus clouds.

No heavy rain appears and the updrafts aren't too wild, so I keep on flying. However, I sneak around the flat bases of some of these big cumulus, remembering stories of pilots who have literally been sucked up inside them. The Pup doesn't have any blind flight instruments and I'm still working on my instrument rating, so this is no place for me and my Pup.

From Flagstaff, it is all downhill to Winslow. I look down into the historic hold of the meteor crator and remember the story of one of our Santa Paula pilots—an airline type, yet—who explored the inside of that crater on a Cessna 150 ferry flight. He couldn't climb back out and busted his bird. I think I can see some of the remaining aluminum, but it's probably all long-gone.

The weather builds up ahead and soon there's lightning in all quadrants. Black cloud cells are everywhere except along the old highway. I sneak into Winslow before one of them moves over the airport. At the gas pumps, we all get wet and the awesomeness of the high desert weather makes it easy for us to stay overnight.

It's time for dinner and we have a choice of excellent Chinese food or a good western steak. Then it's a quick review of the charts and sacktime (Fig. 6-11). TV reception in beautiful downtown Winslow is lousy. Once the amazing sunset has disappeared, it's time to call it a night.

By 7:30 a.m., we're back in the air and headed over the hill toward Gallup. This trip along the Southern edge of the Painted Desert has to be the easiest navigational chore in the country. Just follow the road and the railroad, soak up the great scenery and relax. The hot weather amplifies the smell of gas in the Pup's cockpit and I begin to get a headache.

After 2.1 hours for me, we land at Grants, NM, for fuel. The Pup is burning 4.5 an hour. When we took off from Grants, I struggled into the air without too much trouble, even though the base elevation was 6520 feet and the outside air temperature was

Fig. 6-11. John Peterson studies Sectional charts on the wing of a Bellanca Viking between hops with the Bullpup.

in the mid 70's. However, takeoff was uphill to the west and I had a great deal of trouble getting more than 50 feet in the air—probably because of the symmetrical airfoil on the Pup. I had a good look at the car wrecking area and the city dump west of Grants.

Then we're back down the highway for Albuquerque where we go into the no-radio field at Alameda. Flight time is only 1.05 and we save a great deal of time by staying away from the busy air carrier airport. Even if we'd had radios, we'd have gone into the smaller field.

Mr. Dewey checked weather and the forecast winds are strong enough to throw in the sponge for what remaining daylight we have. We stay overnight. The 7Bar Flying Service socked us $10 per plane per night for hangar space, but there was a possibility of severe weather, so we paid it. The Rio Rancho Inn is close and the TV was better than Winslow.

When you fly a slow airplane without radio and IFR capability, you'd better program your mind to put in a considerable time in motels with a walk around a town that is new to you. The completely unplanned flight of this type would drive a rigid, spit-and-polish type up the wall, but if you program your mind to just one leg of a flight at a time, relax and enjoy it, you'll have more fun and fewer ulcers. At least, that's what my Dad says, and he's a flight surgeon.

By the next morning, I'm thoroughly sick from smelling gas fumes. We scrounge a piece of flex hose from a pool sweeper that the janitor is using at the Rio Rancho Inn and take it to the airport with us. Then it's bailing wire and electrical tape on the airplane to shoot some fresh air into the cockpit. Since both of these ships are in the "experimental" category, we're super legal. After all, Mr. Dewey worked for the FAA for 25 years (See Fig. 6-12).

We head out over the hill, and it is quite a hill, toward Tucumcari, NM. The Bullpup does pretty well at this altitude and the cool air blast from the pool tube is great. I begin to feel better and have hopes that my small breakfast will stay put.

The winds pick up again, and from the wrong direction—so it takes 2.8 hours to reach Tucumcari. Good out-west wide-open spaces with the security blanket of the highway and railroad off my right wing make for great flying.

Mr. Dewey is already filled up and hopefully ready to go, but it doesn't work out that way. The line boy twists the wing tank in the Fairchild with one of those high-powered hoses. It leaks at a pretty good rate, but the best way to go is to empty it and fly only on one tank. This gives the Fairchild a range of only three hours maximum and 300 miles, but we don't want to spend the time to do a field repair. Anyway, it's only a short hop to Dalhart, TX. Both of us have been there before and know the nice people who run the FBO there, Cecil Ingram Flying Service. They help us change a plug on the Fairchild and get us gassed up.

About that time, a flight of new Cessna 150 "nosedraggers" land on a ferry flight to somewhere in the West. I ask them how their trip was going, how the weather was northeast where we're headed, and all that interesting pilot stuff. They couldn't give us the time of day. It seemed that the only thing they had in their minds about our airplanes was that they were in the way at the gas pumps. This made me think about the different attitudes people hold about aviation, the feeling of going cross-country and the great scenery to be absorbed. Probably this crew was out of its element in the isolation of the Dalhart Airport.

It's late in the day, but we give it another go. The new plugs hold and we land in 1.1 hours at Liberal, KS, where Beech makes the Skipper and their other single-engine trainer. Here the LaFonda Motel proved a good place to stop.

In the morning, Liberal is foggy and so is everywhere else. You really don't expect this so far from the oceans, but it's there. We wait and finally that part of Kansas goes VFR. We go only about 15 minutes down the road and the fog, rain and one to one-half mile-visibility advise us both to turn around. Decisions like that are fairly easy. As we return to the airport, we had a good view of the cow barns and high-power lines.

Fig. 6-12. Temporary fix with a flex line from a pool sweeper keeps gas fumes out of the cockpit of the Bullpup. Installation was made in Albuquerque after ferry pilot John Peterson became ill.

After an hour or so, it looks better and we decide to try it again. This time we make it 115 miles to Pratt in an hour and 20 minutes. At Pratt, my Pup drinks a quart of oil and we're off again. Visibility has improved just a little, but we still use main roads for navigation as we head toward Hutchison, KS. I kept the point of my pen on my exact position on the Sectional. One eye was out of the cockpit and the other on my watch so that I would know when to transfer fuel. It takes about 200 strokes on that old wobble pump to empty my six gallons of reserve fuel into the main tank. I could put the Pup in a right slip and squeeze another half gallon out of the auxiliary tank.

Out of Hutchison, I picked up another railroad and then a powerline and followed them toward Emporia, KS, 95 miles away, in an hour and 18 minutes. Mr. Dewey planned for Sedalia, MO, and we were there in 2:05, where we spent a lousy night.

We could hardly wait to get out of Sedalia. The hotel tried to stick us with a conference room for the night when the temperature inside was averaging 102 degrees F. In the morning, it was on toward St. Louis. Mr. Dewey disappears into the murk and I watch this performance again. It's just me and Mr. Buhl once more. I pick up a railroad track heading northeast and it soon parallels a freeway heading into town. The weather becomes marginal, but there are several good landmarks.

I see a weather cell up ahead but it doesn't look too bad. I can still see ahead so I decide to motor on through. The rain fouls up vision on my windshield, so I stick my head over the side to see better. Ouch!

Soon the rain gets heavy and I can no longer see straight ahead or to the sides. I can look only down the little wing root windows and I start looking for an alternate airport.

I find a little place called Fulton and turn around and land. It was nice to see the Fairchild there, too. That tells me that I'm beginning to come up with some of the same answers that apply to a very high-time pilot. Flight time was a very wet 1:05.

We wait for the weather to clear and then go on to St. Charles Airport north of St. Louis proper. Then off again for Casey, IL, 170 miles in two hours, where it's time to call another halt. We walk from the airport to a place called Richard's Farms for a good dinner. We've earned it.

Again we were up early and took off in very marginal conditions to fly over or around Indianapolis and on to New Castle where I find another one of those fun grass strips; 155 miles in 1:50. This is our last stop before delivering the two planes to Mr. Fergus and his museum.

I remember a maze of small country roads and high powerlines. Our destination airport of Union County is about 25 miles northwest of Columbus, OH; in a town called Marysville. Here we met Bob Fergus who had never before seen the Buhl Pup, although he was familiar with the Fairchild. I was sure we had the right airport when I circled overhead and could see a glistening Ryan STA sitting on the ramp. This was the same Ryan that Mr. Dewey had restored and eventually sold to Mr. Fergus.

A trip like this would be lonesome if you did it by yourself. I've been in that situation on a couple of ferry flights. But when you're flying as a team, it's much more enjoyable in so many ways (Fig. 6-13).

You can always come home by airline, but when you're involved in the general airplane business, you just don't make any money with two pilots sleeping in the coach section of an airliner. That's the reason why Mike Dewey had checked around and found a Bellanca 17-31A (N8867V) in South Bend, IN. He bought the ship primarily so that we'd have a profitable trip home, knowing in advance that this turkey was due for an engine overhaul. When Mr. Dewey and I picked it up we drained copious amounts of water from the tanks. Water clogged the fuel injection system and we again visited Casey to get the system cleaned out. On takeoff from Casey, the doggone gear wouldn't come up no matter what we tried, but Mr. Dewey elected to keep on coming west at 120 mph and air blowing around the wheel wells making a frightful noise. With the

Fig. 6-13. The 1931 Buhl Bullpup, flown by John Peterson, and the 1934 Fairchild 22, delivered by Jim Dewey, make a contrast in today's skies as they travel east.

gear down, the Bellanca Viking was even noisier than the Buhl Pup because there was a deep hum over the wheel wells and you could hear the engine valves tick away.

We took 16 hours flying time coming home and drained another half gallon of water out of the old Bellanca tanks when we arrived back at Santa Paula.

It was a bonus for me to work in the same cockpit with Mr. Dewey. Since he is a CFII and FAA designee, I was able to log the entire return trip as dual instruction. We were gone a total of ten days with almost 40 hours in the air covering over 5,000 miles over the ground.

As I'm building up flight time and experience on these ferry trips, I'm planning to go back to Loyola (LA) and get my degree and then get an instructor's rating. Then I want to try for the airlines. I'm in the proper age bracket and I like to fly.

Yes, ferry flying can really be your key to far away places!

Chapter 7

When You Can't Deliver

Not all ferry flights are peaches and cream where you taxi your shiny new ship up to the dealer's door, bound out of the cockpit with a well-satisfied smile, collect your charts and travel bag, turn in your expense account, pick up your check for expenses and go back to the mundane world.

There are those trips that get fouled up in one way or another. Factory delivery delays are predictable. Never head for the factory until you're assured that the new aircraft has been completed and at least test flown. Production promises tend to play hob with your time, temper and expenses.

One time many years ago, I went all the way to the Luscombe factory, when it was in Texas, with a West Coast dealer and two other pilots to pick up three airplanes. We went east in the dealer's four-place Luscombe Sedan. When we arrived, none of the new aircraft were ready and prospects were dim. We finally took the four-placer and all of us returned to the coast together—four wasted days and unproductive expenses.

Even when the new ship is ready, you might not be free and running. Radios don't check out; autopilots don't work properly. The more options in the new ship, usually the more factory check time and the greater chance of a malfunction. Have a fairly firm agreement with your client as to just what he wants done in case of equipment failure at or near the factory. Once you're part-way home, the choice is fairly obvious: get it fixed, or if it is an item

Fig. 7-1. Weather delays are perhaps the single most irritating problem in the ferrying of light aircraft. Here's a sunset near Prescott, AZ, as a Cessna 150 attempts to leave behind virga and thunderstorms. Normally the weather will improve as you travel west.

like a radio or autopilot that you can get along without safely, bring the bird on in.

Weather delays are always part of the game (Fig. 7-1). Be even more cautious about your weather evaluation on a ferry flight than at any other time in your flying career. Remember that you're flying an unfamiliar airplane, usually with strange (to you) avionics, over country that is possibly strange to you, and frequently over extended hours of flight where fatigue sets in. Don't push it.

The mechanical reliability of today's aircraft is excellent. A careful pre-flight is essential, and you hope to find any problems before you leave the factory. More often than not, malfunctions will occur in avionics rather than engine or airframe.

I've had several radios replaced before ever getting off the ground. On the factory delivery of a Cessna 207 to Santiago, Chile, discussed in chapter 10 of this book, it took the replacement of two 300 nav/com units before finding one that would work consistently.

What to do with an in-flight malfunction or emergency depends on so many factors—the mechanical problem itself, your own piloting experience and the amount of time you have in that particular aircraft. *Where* the emergency develops can dictate an off-field landing, looking for the nearest duster strip, or continuing on to a larger town where adequate maintenance is available. The time of day and day of the week can influence your decision. Did you ever try to find a fuel pump for a Piper Cherokee at 3 p.m. on

Memorial Day in Texas? It isn't easy. There are no hard-and-fast rules in any in-flight malfunction and those recounted as personal examples should be taken in the context of the amount of flight experience involved.

In over 50 deliveries since WW II, I've taken the new aircraft to the factory only three times.

The Navion Came Back

The first was with a Ryan Navion (Fig. 7-2) destined for South America. When the cowling came open in Ixtepec, Southern Mexico, there was five-eighth inch chrome molybdenum through-bolt hanging out the side of the 225-hp Continental engine. The bolt had been cotter-keyed and had broken because of loose pyle nuts. Getting that ship, N4916K, partially repaired in the field and returned to San Diego, CA was an arduous, frustrating and expensive process.

After the Ryan factory did extensive rework on this Navion, Swiss ferry pilot Berni Dardel delivered N4916K to Santiago. Dardel reported "a particularly hazardous trip for there are no beaches en route and swamps extend 10 to 30 miles inland."

Fuel Tank Failure in Flight

The second time was when Tom Roberts and I started out from the North American Rockwell factory in Albany, GA with two no-radio "Quail" agricultural single-seaters (Fig. 7-3). It seemed a little unusual that one ship took 18 gallons of fuel during our first stop at Dothan, AL and the other took only 8 gallons. There was a slight fuel leak in N7766V, the ship I was flying. Since we were within an hour of the factory, we telephoned and brought both planes back.

As the flight progressed on the return trip, I noticed more and more fuel leaking and decided to land and take another look. A steady stream of green av/gas was dripping from the trailing edge of the right wing, so we decided to switch to the left tank, punch the quick drain and empty the leading tank.

With one tank empty, the trip back to the factory produced considerable slack fabric over the area of the right fuel tank. When we taxied back to the delivery center, mechanics took one look and began filling out the paperwork for another airplane, N7799V. It seems that a mud dauber had jammed in the fuel vent on the wing tip and the engine-driven fuel pump had actually collapsed the tank in flight. With tubular wing structure going through that tank,

Fig. 7-2. Ryan Navion, similar to this "jet-cooled" model, provided engine problems near Ixtapec, Mexico, that could not be repaired in the field. This Navion, also flown by the author, was at Tuxtla Guttierez, Mexico, during one of the Pan American Road Races.

there was no quick fix, and we headed for the West Coast with a replacement airplane.

On the second start, Tom Roberts and I plodded all the way from the Albany, GA factory to within 50 miles of our destination of Bakersfield, CA. We'd spent an unplanned day and a half in heavy rain at Bryan, TX and we were eager to get the ships delivered and get home. Inbound to Bakersfield, we stopped at Daggett, CA and checked weather with the FSS while on the ground, since neither plane had a radio, which is normal on agricultural aircraft (Fig. 7-4).

Bakersfield was just at minimums with passes reported obscured. However, we taxied out, took off in loose formation and headed for the Gorman Pass. True, it was really socked in. We hooked onto the highway and gave it a good shot, but the clouds hugged the hills to a spot where it became ridiculous. We both rolled into a tight 180 and went back to Fox Field. A phone call to the expectant owner indicated that it might remain IFR for the next day or two. He okayed tying the two ships down at Fox Field, leaving the paperwork and keys with the airport manager and getting back to the Los Angeles area the best way we could. A bus ride and telephone call to a friend to pick us up at the depot concluded that trip.

Back to the Factory for Repair

The third turn-around also involved fuel. The new ship was a Cessna 172 with a seaplane prop headed for Anchorage, AK. Immediately after takeoff, I smelled a little fuel, but the odor soon went away. When we dropped flaps for a landing at Lincoln, NB the fumes increased momentarily. There was a slight red stain down the right side of the fuselage after landing. I had the right quick drain replaced.

There was no odor until the flaps came down again for a landing at Yankton, SD. This time there was a heavy marked stain on the fuselage, so I called Keith Tobin, then Director of Cessna's Fly Away Center. He said that he'd "have to turn me over to a salesman," and I had the feeling that he was trying to tell me something.

I explained the problem and was advised that the Cessna maintenance center in Fargo, ND—right on course—could repair anything that was needed. Then I was given the name and number of the repair shop. It was hot and uncomfortable on this sunny Saturday afternoon in South Dakota and I finally decided to call ahead to make sure that this shop could make repairs and do it in a hurry.

Fig. 7-3. Quail agricultural airplane, with no radio, heads for the West Coast from the North American Rockwell factory in Albany, GA. One fuel tank on this model collapsed in flight (courtesy Tom Roberts).

Luckily, I got the boss on the phone and explained my problem. He assured me that they could fix anything, but he cautioned, "If we have to get any parts from the factory, you'll be here for four days."

After a short gasp for breath, I asked what he'd recommend. The reply was simple: "You're only about 3½ hours from the factory and tomorrow's Sunday. I'd take the ship back."

That's what happened, and we shoved the keys with a brief written description of our problem under Keith Tobin's door before heading for the motel. It was a day and a half later and at least five factory test-hops before maintenance men discovered that a gasket had not been installed inside the right fuel tank filler neck. When the flaps came down, a low-pressure area was formed under the wing skin and fuel blew out of the tank and down the side of the fuselage (Fig. 7-5).

Cessna picked up the tab for motel and meals during the delay but we had less time to spend in Alaska. When we pulled into Fairbanks, AK, eight days and 35 flying hours later, we made the mistake of checking in with the dealer in Anchorage. He re requested that we leave the ship right there because a sales prospect in Fairbanks was interested. The next two days were rainy and miserable, so we saw the middle of Alaska from the train affectionately nicknamed "The Moose Gooser," and then we came home by commercial jet.

Some Fuel Problems Don't Go Back to the Factory

At one time, Aeronca built a rag-wing four-placer in Middletown, OH. It was kind of slow and boxy, but proved popular

Fig. 7-4. Ferry pilot "hams it up" as heavy-rain delays finally cancel the last leg of a ferry flight to Bakersfield, CA. The two agricultural planes were "delivered" at Fox Field to the Airport Manager (courtesy Tom Roberts).

Fig. 7-5. This Cessna 172, photographed at Northway, AK, was returned to the factory to fix a faulty fuel tank. It took the factory a day and a half to find the problem (courtesy Dana Downie).

in its limited way as a bush airplane. I ferried production No. 13 from the factory to Los Angeles after taking a Ryan Navion from San Diego to Reading, PA. That was in the days when two-way ferry flights were possible because of the West Coast production of smaller aircraft (Fig. 7-6).

The Aeronca Sedan's call sign was N1013H. A log book shows seven hours in the air on the first day from the factory to Boonville, MO; then on to the Cessna factory to do a magazine article on a Cessna 170 with Derby Frye. Little did I know that many years later, I'd be the proud owner of one of these four-place Cessna taildraggers.

With 13H, we left Wichita and plodded westbound into the predictable head winds. Radios were not yet production items so landings were made, as they are today in NORDO aircraft, at the smaller outlying airports. It was Tradewinds in Amarillo, and then 40-mph head winds caused us to log 1:26 to Tucumcari and another 3:23 for the 190 miles into Albuquerque, or roughly 57 mph ground speed. Even the trucks going uphill were faster.

We topped off the fuel tanks in Albuquerque and spent the night. On pre-flight the next morning, the tanks were not full and we added an extra 15 gallons before heading out to Gallup (Fig. 7-7). It was not until after the airplane was delivered that we found out we'd been flying an airplane with a built-in malfunction. A

Fig. 7-6. Early-model Aeronca Sedan, production No. 13, is shown in flight. The ship siphoned fuel in flight. Factory telegrams to the pilot were not received because the no-radio airplane was flown down highways and not on flight plan.

factory telegram had been sent to dealer "Hank" Coffin at the old Vail Field, advising that the fuel-venting system was siphoning out gas in flight. We knew that it was venting on the ground from the fuel that had disappeared overnight and had paid particular attention to fuel consumption on the tail end of the flight to Gallup, Prescott, Lake Havasu and Twenty-Nine Palms.

Remedy for the problem was simple once the facts were available. The factory had installed all the fuel caps on the first 13 airplanes backward (remember Murphy's Law?) and every ferry pilot had been notified through the dealer of the problem involved. We hadn't telephoned in, since Hank Coffin was in no great hurry to get the new ship. Our fuel bill was mighty high on that trip.

Most Malfunctions are Nav/Com

More often than not, minor malfunctions don't require a long return to the factory. For example:

It was a chilly morning in Auburn, AL, and it took considerable alcohol and rubbing to get the frost off the wings of the Grumman Tiger just two days before Christmas. The early morning was crystal-clear as I lifted off alone in N28558 and headed for Memphis for fuel and breakfast. Again, it didn't work out that way.

There were scattered, light snow showers, but the outside temperature was well below freezing, so icing was no problem. I worked Memphis Approach from the Holly Springs VOR, planning a landing at the big international terminal. About that time, the nav half of my single radio rolled over and died and the communications section became intermittent. International went below VFR minimums, and I was able to get the idea across to Approach that I'd like vectors to West Memphis. No problem—I crossed the Mississippi River and finally received a clearance to land. There was no coffee shop or radio shop! However, the FBO assured that the small General Dewill Spain Airport, about 10 miles northeast, had both a fine radio shop and food. They did, and I cooled my heels (better to say that I warmed them) while a sharp technician dug down through four layers of Narco circuit boards to find a loose piece of solder. Of course, the new equipment was still under warranty, so payment for the avionics repair was taken care of between the manufacturer and the shop. The dealer, Joe Geiger's Performance Aircraft in Long Beach, CA, didn't have to send any checks.

Carbon Monoxide Is Sneaky

When Cessna first introduced its "strutless wonder," the Cardinal, all the aviation press attended. The managing editor for AOPA Pilot wrote about the airplane from the factory standpoint, and I ferried the first one, N2214Y, to the West Coast. Our articles ran back-to-back in the magazine. The solo delivery west was uneventful up until the final leg where a minor problem developed which could have had very serious results.

I had filed from Ely, NV, to Reno with a 2 p.m. local-time takeoff. As the turbulence flattened out a little, I tried to plan ahead. The charts showed another 195 miles from Reno to San Francisco and some very simple arithmetic proved that if I landed at Reno, it would be dark by the time I reached San Francisco.

So, with that 5.8-hour (no reserve) fuel range, I called Lovelock Radio for the latest bay area weather and a forecast for two hours later. Surprisingly, there was no fog and only broken clouds en route over Lake Tahoe. I asked the FAA/FSS to notify Reno since I couldn't yet reach them and extend my flight plan. I detailed the new airplane and explained the 177/Cardinal relationship. I'd filed as a C-177 because I didn't want to explain to every FSS what a Cardinal was. The next pilot through this mountainous route probably won't have that trouble because the FAA has already been briefed.

I climbed back up to 10,500 to go over Echo Summit south of Lake Tahoe and added some cabin heat. To that point, 14Y had

Fig. 7-7. Close-up of fuel vent of the Aeronca Sedan on the ground during a dust storm at Albuquerque, NM. Fuel can be seen venting from the drain, even though the aircraft is on the ground.

been a completely regal machine, befitting her name. However, a delivery flight in any new airplane without at least a minor malfunction to take your mind off the scenery is not always the case.

As I pulled out on the cabin heat, the selector stuck in the full "on" position. A factory press release advises, "A high-volume cabin heating system insures adequate heat during cold weather operation." I can assure you that this statement is 100 percent correct. After I topped the Sierra Nevadas and began a letdown south of Sacramento, I had every cabin vent in the airplane wide open. It was hot, but neither dangerous nor unbearable. At 4500 feet, I leveled off and slowed up to the allowable 120 mph indicated before cranking open the air-scoop windows at the forward edge of each door. These vents cooled things down considerably (Fig. 7-8).

Sure, I could have landed at Sacramento and had the heater control fixed, but it was only another 75 miles to my destination. Then Mr. Bergines could have his own mechanics solve the minor malfunction.

So, we cruised slowly into SFO—warm as toast and enjoying every moment of the then-calm air. Our total flight time from Ely was an even four hours. Yet each fuel gauge still read over one-fourth tank.

I returned to Southern California on a jet commuter with a driving headache that was unusual for me. Even the following day the headache persisted and I wrote it off to travel and editorial deadlines. It wasn't until several weeks later that the Western

Fig. 7-8. Early-production Cessna Cardinal on the ground for fuel at La Junta, CO, prior to a solo flight over the Rocky Mountains and on into San Francisco. On the last leg of this flight, a faulty heater muff put carbon monoxide fumes in the cabin.

Factory Representative told me what had really happened. The exhaust stack on the new engine had a crack that fed into the heater muff. With the heater on, I had absorbed enough carbon monoxide to have the memory linger on. Fortunately, the cabin became so insufferably hot that I used the wide-open, air-scoop-type vent and had ample fresh air to mix with the fumes. Otherwise, my "terminal airport" of San Francisco International might well have been changed to some snowy rock in the Sierra Nevadas.

It's usually a question of economics with a long weather or mechanical delay. If you have a serious engine problem, it is frequently desirable to get the ship safely tied down at an airport or in a hangar, and let the owner make his own arrangements for the expensive repairs. Make a phone call to the owner and list your options so that he has no surprises. Be sure that the plane is protected the best way possible before leaving it. If hangar space is available for a day or two, use it. Then you get home the best way you can.

Chapter 8

The Odd And The Rare

When you ferry over any extended period of time, you're sure to climb into some rare birds—one-of-a-kind models, aircraft with very limited production, homebuilts, floats and imports.

The following are ferry experiences that cover the whole world and some nooks and crannies within the United States.

Ferry Flight to Nowhere: First Homebuilt Around the World

Departure from Oshkosh—destination, Oshkosh. Total time: 61 days; total flight time: 171.5 hours. Homebuilder Don Taylor had all the problems associated with a long-range ferry flight. His goal was to do something that had never been done before—fly a homebuilt around the world.

Actually, the record-breaking trip included two attempts—the first when Taylor's T-18 came home by boat from Kurshiro, Japan.

The following account was published in AOPA Pilot magazine.

Few people get into the record book the easy way, and tenacious Don Taylor is no exception. Eleven years earlier, shortly after he had retired from the U.S. Air Force as a Lieutenant Colonel, Taylor decided that he wanted to make a flight that would go in the record book. One way, back in 1965, was to be the first man to fly a homebuilt airplane across the Atlantic.

First he had to learn about small airplane construction and small engine theory, design and construction. That took eight months at the Spartan School of Aeronautics in Tulsa and earned Taylor a powerplant rating to go with his 10,000-plus hours of flight time.

He then spent six months looking at all available homebuilt designs and soon had his choice narrowed down to a Midget Mustang or Thorp's T-18. He decided on John Thorp's T-18, a tiny taildragger, and went to nearby Burbank, CA, to talk with the veteran designer.

"Can you put enough fuel in the T-18 to fly 2,000 miles?" Taylor asked.

"Yes," replied Thorp, who then asked why. When Taylor explained his goal, Thorp tried everything he could to discourage the would-be record-breaker.

"I had to push him all the way and really convince him on the project before he would sell me a set of plans," explained Taylor. "Even then he was against the idea because he thought I might drop in the drink. However, once he saw I was going to try anyhow, he modified the plans so that he felt I could take off safely with a 100% overload."

Taylor copied all the T-18 templates and made many of the sub-assemblies in Thorp's small shop. The strictly stock 180-hp Lycoming 0-360-A2A was one of a group that Thorp had purchased and later sold at cost to T-18 builders. He completed assembly of his T-18 in a garage adjoining his private Ernst Airport, just 14 air-miles from Hemet. Call sign for the T-18 is N455DT for Thorp's print number on the record airplane, plus Taylor's initials (Fig. 8-1).

The first 50 hours of "shake-down time," required by the FAA on any new homebuilt, were flown by Taylor from his front yard airport. Then came the first, long cross-country flight, only 1700 miles from Hemet to Oshkosh in 12 hours, 43 minutes.

Modifications and test flights were spread out over several months. Taylor goes on the personnal theory that, "if an engine will go 200 hours, then it'll go 2000."

The wiley Taylor named his T-18 "Victoria" because it coincides with the only one of Magellan's fleet to complete the first around-the-world trip back in 1622. The name Victoria was popular with the British and kept both the Spanish and Portuguese happy. It was even useful in the Philippines due to their early Spanish history. "Victoria 76" was adopted later to reflect the bicentennial year.

Taylor didn't make it on his first attempt, but he certainly gave it a good try. He admits that the first attempt was the most

Fig. 8-1. Round-the-world flyer Don Taylor was a USAF fighter pilot, so close-in formation with his T-18 was like old home week.

nerve-racking in all respects. Although challenging, he admits that "number two didn't seem so bad." His first record attempt was in August, 1973 and ended 63 days later on a 3500-foot runway, still under construction, at Kurshiro, Japan. Arctic winter weather forced him to turn back. He disassembled his tiny T-18 and shipped it home by freighter. This cost him $2200, including $800 for the crate that Japanese workmen built around N455DT in one day. Today, the wood from this crate forms the structure of a hangar in front of his trailer home near Hemet.

On his first attempt, Taylor felt that he had to decide between survival gear and oxygen. There wasn't room for both, so the O_2 bottle stayed home. Crossing the Alps, he had to climb to 17,500 feet to top icy clouds before shooting an ADF approach to Rome's Ciampino Airport. Because of anoxia, Raylor admits making two incorrect heading changes during the early portion of this letdown.

Taylor found that international red tape, not fuel capacity, limited his daily range. At the last minute, he was denied landing permission at Baghdad. All American aircraft were prohibited over Iraq, so he detoured over the lonesome Dasht-E-Kavir desert of Iran, climbing to 15,000 feet to clear a sandstorm. "Worst desert in the world. Visibility nil at 15,000," said his flight log. It was eight days more to Calcutta on the first trip.

En route, he found that his planned fuel stop at Raipur, India, had no av/gas, just jet fuel. Therefore the former fighter pilot used the first of a series of ploys, filing VFR to Raipur, declaring a fuel shortage en route and landing at Nagput, some 100 miles distant. He refueled and was back in the air before governmental complaints developed. This same procedure, while less than perfect, actually made it possible for him to complete his goal three years later (Fig. 8-2).

Fig. 8-2. Don Taylor and his wife, Lois, look at a portion of the world covered in the first successful flight of a homebuilt to circumvent the globe.

After his first attempt, tenacious Taylor really wasn't through, even though he told his many flying friends, "No, I wouldn't do it again. Well, not unless I had enough financing for good food and rest as well as some refinements on the ariplane. I'd certainly add a cockpit heater and oxygen for safety. I'd want a more comfortable seat. I'd replace the engine cowling for better cooling and speed, update the radios and add a more powerful transmitter." The Narco Mark 12A that made both flights came out of a wrecked Bonanza over 10 years ago. "Then, I'd remove the baggage fuel tank and add 10 gallons under the seat where it has a more forward CG."

At that time, Taylor said, "The mid-Pacific route is out. A 2200-pound Class C-1-b airplane just won't carry enough fuel to make it. Some of my friends wise-cracked that I could complete my hangar with the crate from the next trans-Pacific shipment."

Just three years later, Taylor and his T-18 were back to try it again. He'd made all the changes required in the T-18, aided by both "spiritual and practical help from local EAA chapters."

Now an "old-timer" at flying his T-18 to the far corners of the world, Taylor had his international paper work fairly well in hand. His application for landing permission stated in part, "These countries and their airports are the best and most convenient in which to land, refuel and rest. They were selected to expedite the flight. No passengers and no cargo will be carried on this flight. All stops are for crew rest and refueling only."

The paint job on Taylor's Thorp is deliberately gaudy so that it can be seen on the desert, against foiliage, over water, snow or ice. Basic elements of the $375 survival kit that Taylor has never opened include a two-man raft, flares, thermal blanket, fishing gear, etc.

A 36-cubic-foot high-pressure oxygen bottle was strapped securely in the right seat. Taylor used oxygen on only three of the 37 flights of his second, successful attempt. "As every pilot knows, it's foolhardy to try precision IFR flying at altitudes above 12,000 feet, even if you're a young tiger. I tried it without oxygen on the first trip and had some hazy spots at 17,500 feet over the Alps. I had no such problems on the second flight and ran my oxygen tank dry on only one occasion."

According to Taylor, one of the truly inspirational aspects of the entire project was the cooperation and dedication of fellow homebuilders all over the world. He admits freely that he is not the most adept of aircraft craftsmen. Thorp has described his workmanship as "safe but not outstanding." Nearby EAA members developed a camaraderie in helping Taylor who painted a list of about twenty names on the top of the fuselage "for the most persistent helpers, but there isn't room on the whole airplane to list everyone who pitched in." (See Fig. 8-3).

The engine is fed directly from the standard Thorp 30-gallon gravity feed tank forward of the instrument panel. Then the engine-driven fuel pump takes fuel from any of the other tanks when the selector is changed. Taylor attempted to always take off and land on his main tank with fuel that he was sure about. He used pick-up fuel of sometimes-questionable vintage in wing and seat tanks.

N445DR did not carry a reserve oil tank. Taylor said that he used less than one quart of oil in 10 hours on long, hot flights and only a quart in 15 hours in the Artic.

Fuel availability was a continuing problem. In Bangkok, the Thai Air Force finally drove up in an old truck with 1,000 gallons in the tank. "All I knew was that it was purple," said Taylor philosophically. On the Pacific Islands of Truk and Ponape there were thousands of gallons of jet fuel, but the av/gas came out of old 55-gallon drums via a wobble pump. There was no practical way to

Fig. 8-3. The instrument panel of Don Taylor's T-18 is anything but plush. Sticky-tape and typewritten placards are much in evidence. Add-on mounts for instruments show the cramped space in the small cockpit. However, there is enough nav/com equipment here to get the job done.

test the age or deterioration of the fuel but to start up the engine and try it.

Taylor's "impossible" mid-Pacific crossing was relatively uneventful as far as Wake Island, "where the U.S. Day begins." His next planned landing, and the only way the mid-Pacific route would work for a plane with the range of the T-18, was a fueling stop at Midway Island; but a landing approval was never received. Perhaps it was because the U.S. Navy didn't want to go out and look for him if he turned up missing.

Since Taylor didn't want to send his airplane home a second time on one of the few freighters to stop at Wake, he took on 120 gallons of fuel and filed a VFR flight plan to Lahui, the north island of Hawaii, some 2400 miles distant.

"It was a funny leg," admitted Taylor. "I had the headwinds at 5000 feet that you normally get out there. There just wasn't any dry place to go except Midway, so I ammended my flight plan after 8½ hours and made it into the big U.S. Navy base with no problem."

Since he had landed without prior approval, Taylor expected a dressing down. However, he hadn't expected the Navy to impound his T-18. Taylor hadn't spent 22 years in the military and been a fighter pilot for nothing. He called on his reserves, members of the EAA, flying friends of congressmen and everyone else to obtain permission to fly out of Midway.

No one is really sure of the chain of events, but it took approval from the Chief of Naval Operations at the Pentagon to get Taylor a full load of fuel, a $100 fine and permission to take off for Adak. He waited an additional three days—16 days total on Midway—for that perfect weather forecast. "If I'd waited another week, I'd have been in a hell of a mess," said Taylor candidly. "The days kept getting shorter and the weather predictably worse."

"That dawn takeoff from Midway was one of the greatest experiences of all my flying days," reminisced Taylor. "Weather had said it was go now or not at all. I'd overstayed my welcome and really wanted to go. Yet, this was somewhat of a test hop since I'd never before flown the T-18 crammed full of fuel. On all the other hops, the 15-gallon extra tank had been carried empty in the baggage compartment. Out of Midway, it was in the right seat and full.

For the first time in his second 'round-the-world attempt, he had taken off with full fuel—145 gallons. His 20-foot, 10-inch Thorp

T-18 had an empty weight of 1050 pounds and his gross at takeoff was 2150, more than double the empty weight of the airplane. Normal takeoff weight for the T-18 is 1500 pounds.

"I climbed out at 500 fpm on 2400 rpm. All these flights were made with a fixed-pitch prop to save 20 pounds and $1200.

"I wanted to have just enough sunrise to see what I was doing on takeoff at Midway," explained Taylor. "If I waited any longer, I'd run out of daylight 1650 northward, and Adak isn't the biggest island in the world. If I missed it for any reason, I planned to land on the first island of the Aleutian Chain that I could find before dark.

"I remember looking back over my left shoulder after heading out from Midway and watching that island go out of sight. I almost turned around and went back, but after all the effort and the help that had gone into the trip, I kept on going."

"When things settled down, I tuned in the HF receiver and heard the Anchorage FAA crew calling me. They called every half hour for the entire trip. I met two of the operators three days later. Great people."

"I'd been in Adak just once before in the right seat of an Air Force C-47 back around 1954," said Taylor. "All I could remember was that the approach was mean and I wanted to get in before dark, so I pushed my rpm up slightly to 2325, picked up that ever-loving ADF 200 miles out and landed 30 minutes ahead of my ETA with 45 minutes of daylight and three hours of fuel remaining."

From Adak, it was all "downhill" along the surface of the globe to Cold Bay, 650 statute miles away, to Anchorage, Whitehorse, Fort St. John, Edmonton, Minot and back to Oshkosh, where he had departed during the EAA's bicentennial fly-in.

However, he did make an unscheduled second landing at Edmonton. Some 20 minutes after takeoff, his dependable Lycoming began to shake and shudder. A mag check showed a drop of "plenty" on each side, so Taylor made a 180 and returned to the airport. He and a mechanic pulled the cowling and spark plugs. The bottom plugs were completely fouled and the top ones weren't so good, so Taylor threw them all away and put in an almost-new set from a nearby aircraft. The plugs cost Taylor $30 and are still in his T-18.

"I really should have changed plugs in Anchorage, but didn't because I was too cheap," he admitted with candor.

Two days later, he was back in Oshkosh where his wife, Lois, and EAA members met him. He celebrated his 58th birthday there after a 61-day journey of 24,877 miles in 171.5 hours, and entered the record book as the first man to fly around the world in a homebuilt (Fig. 8-4).

It was really a ferry flight to nowhere except an entry in the record book and the personal satisfaction of completing an 11-year dream.

To Paris with Turbo DC-3: They Only Built One

Airline Captain, race pilot and motion picture pilot Clay Lacy is not really a ferry pilot. However, he did take the assignment from owner-designer Jack Conroy to act as captain when the one and only Turbo DC-3 went to the Paris Air Show. In addition to Lacy and Conroy, Herman "Fish" Salmon shared in the piloting chores.

I went along for the ride from Los Angeles, did a part of the flying and some reporting. Coming home, I was able to "dead head" back across the Atlantic in a U.S. Coast Guard C-130 that had a display helicopter in the big cabin. Then I ferried what looked like a very small two-place American Yankee west from Cleveland.

The design of the "Turbo 3" was yet another effort by Conroy to salvage the Rolls Royce Dart Mark 510 engines and their propellers for use in a time-proven airframe. The engines and props were readily available and the price was right, as the British Viscount transports with their rigid wing structure were limited to some 15,000 hours. Therefore, there was an ample supply of reliable engines and props that Conroy hoped to mate to the popular workhorse of the DC-3 (Fig. 8-5).

Conroy was no newcomer to the big airplane-modification business. He built the "Guppy" series of huge transports to transport experimental space rockets around the airframe of a Lockheed Electra. Years later, he used the identical airframe of the "Turbo 3," N4700C, to become a three-engined turboprop patrol and bush plane. Conroy is certainly tenacious.

Actually, Conroy's model was not the one-and-only turboprop version. At the conclusion of our ferry flight, I talked with British test pilot Cliff Rogers in Paris. He explained that three "Dakotas" (British designation of the C-47/DC-3) were converted to turboprops in 1949. In fact, the first of these was used as a full-scale flying test bed for the original Viscount turbo-prop combination. The remaining experimental Dakotas were used on freight runs and to check out British European Airways pilots who were sched-

Fig. 8-4. Don and Lois Taylor live in a comfortable trailer home on a dirt strip near Hemet, CA. The partially-completed hangar is made from the packing crate that Taylor used to ship his homebuilt back to the United States after his first unsuccessful trip.

Fig. 8-5. Jack Conroy's Turbo DC-3 on the ramp in Santa Barbara, CA, prior to a ferry flight to the Paris Air Show. The venerable transport was powered with two turbo-prop powerplants removed from a British Viscount.

uled to transition to Viscounts. It was reported that these ships later flew coal on the Berlin Air Lift, but Rogers had no proof of this activity.

N4700C had been used by Conroy for several years before we went to Paris. When one of the standard R-1830 engines came apart in Baja, CA, Conroy brought the DC-3 back with a costly replacement engine and decided to convert to turboprops.

The engines were two of the lowest-time powerplants (3000 hours each) available on United Airlines runout Viscounts. Engineering studies dictated that the light-weight engines and props would have to go far forward of the original powerplants and extended nacelles were built. Because of the higher fuel consumption of the turbo engines, the entire wing was sealed off and "wetted" to double the fuel capacity to about 1600 gallons. Average fuel consumption was 100 gph per engine which gave a dry-tank endurance of eight hours. At 15,000 feet, the Turbo 3 would true-out at 235 mph.

As with so many ferry flights, this Paris trip was a last-minute operation. The new conversion had only three hours of flight testing when we departed from Los Angeles. Lacy checked the engine feathering and all emergency systems just before we landed at Montreal. We spent a couple of days there while flight mechanic Louis Salazar plugged hydraulic leaks and worked on the fuel crossfeed system and cabin heater problems (Fig. 8-6).

This was my first trip over the North Atlantic, although I'd made the Africa-Azores-Bermuda trip twice before. I was ex-

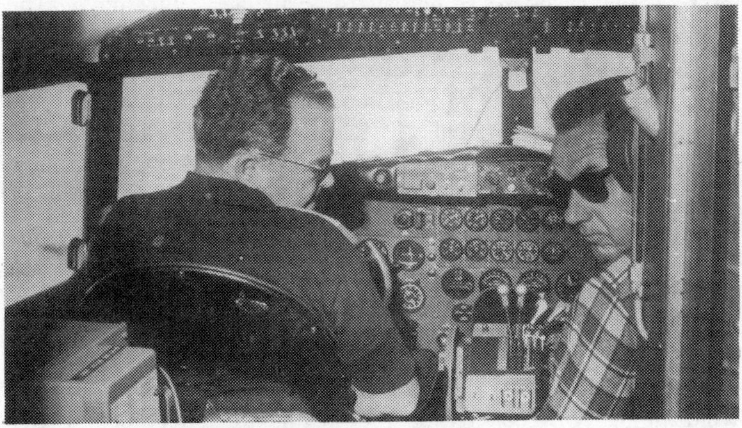

Fig. 8-6. Jack Conroy holds down the pilot's seat while Clay Lacy rides "shotgun." Note the simplified power levers in the Turbo 3.

tremely happy to have the flight experience of Lacy, Salmon and Conroy in the cockpit. At that time, Lacy had nearly 20,000 hours of airline flying; "Fish" had just retired from Lockheed after 29 years of test flying; while Conroy was an ex-B-17 pilot who had built up 20,000 hours in the rough-and-tumble world of "non-sked" airline flying following WW II.

On a test-hop at Montreal before leaping off over the water, Lacy and I went out alone. I found out very rapidly that the Turbo 3 was a different breed of cat from the military C-47. To begin with, the propellers turned "the wrong way" and torque on takeoff was to the right! There were only two power levers in the cockpit—one for fuel flow and one for prop control. Takeoff was at 14,500 rpm with a 10:1 reduction to the propellers. An overspeed to 17,000 rpm was permitted for 20 seconds (Fig. 8-7).

It's a liberal education for any pilot to work on pre-flight planning, weather evaluation and a hard look at escape routes on a flight like the trip across the North Atlantic. Lacy, in particular, had an uncanny knack of picking weather alternates that were well within range, but that didn't require completely retracing the route.

Our first planned stop out of Montreal was Goose Bay, Labrador, but as so often happens in ferry flying, we didn't get there the first time out. We had covered about half of this four-hour flight when Goose Bay weather slid down to 500 feet and three-quarter miles with snow showers. The field was supposed to go below IFR minimums within the next hour. So we circled the colorful French Quebec town of Sept Isles, talked with the tower, found out that jet fuel and accommodations were available, and landed. The Chinese food in the French restaurant that night was excellent!

Goose Bay came up to acceptable weather minimums the next day and we came in with four miles visibility, a 2500-foot ceiling and only a light rain. "Fish" Salmon happened to draw the left seat on this leg and Lacy slapped a map across the windshield just as we broke out of the clouds.

"Just to keep you honest, take her right down to minimums," said our Captain with a grin, "that's two hundred and a half." And Salmon went ahead to prove that he was a proficient instrument pilot as well as a fine test pilot.

During this letdown we listened to the incessant banging as ice spun free of the props and slamed ito the fuselage ahead of the windshield. With conventional powerplants and stubby engine mounts, this shower of ice cubes usually hits aft of the cockpit windows.

Fig. 8-7. French language airport signs are on the Montreal Airport as the Turbo 3 taxies in the background.

We fueled, checked weather, and planned to get right into the air for Sondrestrom, Greenland. This old WW II refueling base has a one-way 11,000-foot runway with a steep fjord at the far end of the field. Just before start-up, the reported weather went to wind gusting over 40 knots, visibility down to one and a half miles with blowing sand and several other unstable indicators. Lacy cancelled and we headed for the Bachelor Officers' Quarters since most of us aboard had been military pilots at one time or another. At that time, the only nearby civilian hotel was in Happy Valley. We stayed on the base.

During our forced layover, we visited with U.S. Air Force briefing officers who normally handle all North Atlantic ferry flights. If you ever have the chance to fly this route, I'd strongly recommend a layover here to pick up an on-site briefing of airports, navaids, radio facilities and search-and-rescue procedures. To listen to this group talk, our AiResearch life raft and salt-water-activated Garrett ELT were really not all that useful. The briefers indicated that the temperature on Greenland's ice cap can be a mild -30 degrees F.; but with a 30-knot wind, the chill factor comes to -82 degrees F., and you'll freeze in 30 seconds without protective clothing.

We spent two nights in Goose Bay's comfortable BOQ before there was a steady improvement in weather. We found out that there were two emergency ice runways, "Sea Bass" and "Sob Story" near our route over the 8000-foot-high ice cap east of Sondrestrom. Greenland and Iceland radars were called "Big Gun" and "Ink Spot." At that time, weather ships "Alpha" and "Bravo" were on station for nav/com assistance.

Our take-off weight from "Goose" was 33,500 pounds, compared with the 25,000-pound limit imposed on the DC-3 in commercial service. By contrast, the DC-3 I'd taken out of Long Beach for the South Pacific lifted off at 31,500 pounds.

"Orphan Annie," the Goose Bay radar, talked with us for more than an hour and then relayed our position to a PanAm jet when our L/F radio refused to do its job. Another hour out we heard that Sondrestrom weather was again headed for the bottom of the fjord with four miles and light fog. There was scurrying both in the cockpit and in the cabin as additional charts came out. We could either divert a long way north to Frobisher Bay or continue toward Sondrestrom with the ice runway beside the "Seabass" Dewline station as an acceptable alternate. Frankly, I would have found the Dewline station interesting and had the faith in our cockpit crew not to sweat a landing there (Fig. 8-8).

Weather at Sondrestrom held at 2000 feet and Conroy, who was in the left seat on this leg, broke out following one of the best ground control approroach radar approaches I've ever seen. The crews at Sondrestrom have to be good to get you down the middle of that narrow fjord and into position where a next-to-impossible go-around is not required.

We taxied the bright yellow and orange DC-3 up beside the neat little SAS airline terminal, fueled (Fig. 8-9) and decided to spend a night—short nights in this latitude in the summer—at the tidy airline hotel that was established by SAS to handle weather or maintenance delays on early North Atlantic flights. The increased

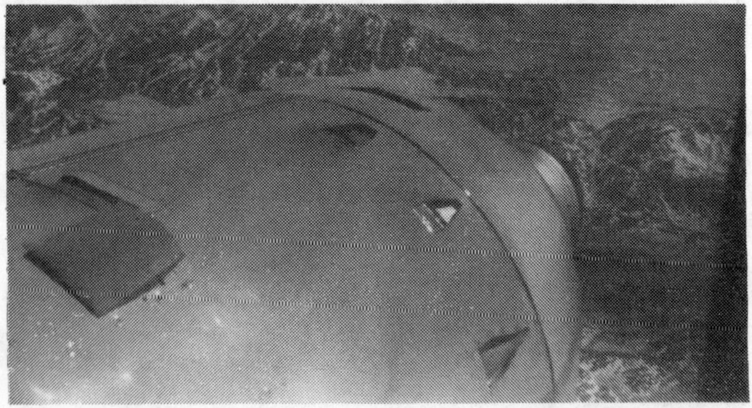

Fig. 8-8. Uninviting mountains and ice floes clog the North Atlantic. This view from the cockpit of the Turbo 3 reminds all on board that this is certainly no place for a forced landing.

Fig. 8-9. The Turbo 3 is fueled at the Sondrestrom Airport. An SAS Airline passenger terminal and small hotel can be seen in the background. It is raining, something not uncommon at this airport.

Fig. 8-10. Descriptive signs point out where the remainder of the world is in relation to Sondrestrom, as the Turbo 3 is fueled. Military ramp and hangars show in the background.

dependability of high-flying passenger jets has now made this stop used only on a rare emergency.

Again, the food was good. Ferry pilots who are forced to delay by weather or mechanical problems frequently spend their time and expense money in sampling the best of local foods, and the SAS Hotel in Greenland was no exception. Outside the cafe and hotel is an unusual signboard with the direction and air-mile distance to some of the larger air capitols of the world. This is a gentle reminder that you're on the ground in one of the more isolated spots on the face of this world (Fig. 8-10).

The sun was shining brightly the next morning. We were able to take a few good color photos as a remembrance before taking off downhill, downwind and down-fjord, circling back over the glaciers and the ice pack of Greenland. We talked with "Sea Bass" and passed this Dewline station close enough to see the radar domes and markers for the ice runway. This would have made a really interesting emergency stop, but we might have waited for days to have fuel flown in had we landed. We droned along over the ice cap toward Keflavik.

Off our right wing we saw the gravel strip at Kulusug ("Big Gun") where almost all the Denmark Straight was under ice. You've really got to admire the people who sign on (or are assigned) to do a full tour of duty in these really remote corners of the world.

Our cabin heater went out and the call sign "Zero-Zero-Charlie" was most appropriate. Everyone wrapped up in whatever topcoats were available and the most popular spot on the flight deck was just behind the co-pilot where the radio deck and hydraulic reservoir generated a little heat (Fig. 8-11).

The airport at Keflavik was neat as a pin and fueling was efficient. Too bad we didn't have a delay here and a chance to visit part of Iceland. Maybe next time.

We took three hours and 45 minutes to get to Prestick, Scotland, and spent the night in a private home with rooms to rent near the airport. Since my father had come from Perth, Scotland, I was looking forward to my first glimpse of this part of the world. No such luck—we were IFR at 1000 feet and on-top all the way across Scotland, England and the Channel. Conroy broke out below very wet clouds at the LeBourget Airport with two miles visibility.

The sunburst paint job on Conroy's Turbo Three was an eye-catcher for the Paris Air Show. Interested throngs inspected the ship and talked of performance, but the project didn't sell as expected. Perhaps in a later environment, where the quiet engine of the turboprop becomes increasingly important to airport neighbors, such a package may again seem worthwhile (Fig. 8-12).

Fig. 8-11. Herman "Fish" Salmon, former Lockheed test pilot, rechecks navigation while crossing the North Atlantic in Jack Conroy's DC-3.

Watching the U.S. Coast Guard C-130 crew drone westbound from Paris to Andrews Air Force Base in Washington, DC, was almost anticlimactic. Here was a proven four-engine turboprop with all the military nav/com equipment available and a high-time flight crew. We stopped for fuel, and since the crew was overdue for a rest period, spent the night at a military base near St. John, Newfoundland. Since I was the only civilian aboard, I suddenly was made a "guest" Coast Guard Lt. Commander by our aircraft commander rather than battling the papermill to get a civilian into the BOQ and officers' mess.

Yes, pilots who ferry airplanes can expect to get into some of the most unusual situations!

Waterports—A Whole New Ballgame

Float plane deliveries present unusual problems. Unless you have super-expensive amphibious floats, you'll be working a whole new package of waterports. Clare J. Wesley, sales manager for St. Cloud Aviation Inc., in Minnesota, has ferried a number of float planes from his base to Alaska. He says that the trip is more interesting, but not more spectacular than most other areas of North America.

"If anything, since the route I fly through Central Canada is generally blessed with good summer weather, I find it easier to fly than in the Southern United States. Air is generally smooth. Winds are not a significant factor and the hours of daylight are a blessing.

"My routing from Minnesota to Alaska takes me through a border clearing station—Baudette or International Falls; then to The Pas, Manitoba, La Ronge, Buffalo Narrows, Peace River, Fort St. John, Fort Nelson, Watson Lake, White Horse, Northway and Anchorage. An alternate routing carries me near Fort MacMurray to Fort Nelson, saving some mileage.

"Refueling facilities are adequate throughout the route; however, one can expect some "do-it-yourself" treatment. Occasionally, you run into the problem of coverage, even during normal daylight hours. Canadian Notam service on fuel availability is non-existent from my experiences.

"During the peak floatplane season, mooring and docking facilities become critical because of the many planes concentrating at the major refueling points. You learn to have plenty of rope and to keep the paddle readily available. Even with this problem, I prefer to use the major facilities because of better flight briefing facilities, weather reporting, housing, etc."

Fig. 8-12. The Turbo DC-3 is parked on the display ramp at the Paris Air Show with several Russian transports visible in the background.

Two St. Cloud employees delivered a Super Cubfloats to Alaska. A report on this ferry flight was written by Rhonda Clayton and was first printed in Saint Cloud Aviation News. Rhonda and Steve Champa made this particular delivery flight in the month of August:

On August 21, Steve Champa and I left from Park Rapids, MN, in a Super Cub on floats, bound for Kenai (Fig. 8-13). Flying to Alaska from the "lower 48" in a small plane was the adventure of a lifetime for us. Clare Wesley had asked Steve if he wanted to deliver the seaplane to a customer in Alaska and he got an immediate "Yes." I figured Steve needed a copilot, so I volunteered myself.

We left with a minimum of baggage—two five-gallon cans of fuel, which were to be used in case of emergency; survival gear; cameras; four Wac charts, and 17 Sectionals.

We cleared customs in Baudette for our first stop. Next was Winnipeg and on to The Pas. I wasn't quite prepared for the extreme contrasts of terrain, weather or lack of human habitation and landmarks that we encountered in NW Canada. As far as navigation went, it was back to basics. Our compass started playing games with us early on the trip. Mountain ranges made it difficult for small aircraft to depend on VFR radio navigation and due to cloud layers in the valleys we were forced to fly at low altitudes.

We saw some of the most breathtaking scenery. Flying from The Pas to LaRonge was gorgeous. Our ground speed checks indicated we were traveling at the tremendous speed of 120-130 mph. We didn't encounter the severe turbulence of winds that we were warned about. But we were in and out of rainshowers, which brought us many beautiful rainbows when the sun came out. Steve made a lot of "glassy water" landings which were nearly perfect. By the end of our trip he was a master at them. The further northwest we flew, the longer the daylight hours became.

I gave Steve quite a scare at Lesser Slave Lake. We needed fuel but didn't know where to get it. Steve said he'd walk to the airport and I should stay with the plane. The plane began to drift from shore. I let it drift and float while I waited for Steve to return. Soon I had convinced myself that Steve and the fuel truck were on the other side by the channel. I climbed from the back seat to the front; I felt excited. I had not step taxied for three years since I got my float rating. I got up on the step with a lot of back pressure, but I forgot one important item—*throttle*! I didn't pull back on the power soon enough and the Cub took off. I did a dandy landing. I wished Clare Wesley, my floatplane instructor, could have seen it. I step taxied correctly this time over to the channel, but no Steve. Meanwhile, he had come back and found us gone—simply vanished. Much to our relief we did find each other again.

The hardest part of our trip was the Ft. Nelson to Watson Lake leg—about 300 miles. Near Muncho Lake the Alaska Highway makes a right-angle turn, which we missed. We followed an old mining road and found ourselves lost. Our plan was to head west over the Liard River until it joined the AlCan Highway. We had mistaken the Toad River for the Liard River. When we found Muncho Lake with the beautiful Alaskan Highway beside it, Steve landed, put in the extra 10 gallons of fuel and headed for Watson Lake, where we spent the night. Fog kept us from leaving until 11 a.m. On the last morning of our trip we came across some of the most beautiful scenery we'd seen so far; glaciers and snow covered mountains—and we saw it all from a bird's-eye view.

Jack, one of the FBO's at Lake Hood in Anchorage, marveled at our long trip in a floatplane since we'd never made the trip before.

Steve and I shook hands with the owner, gave him his plane, and then we shook hands with each other for delivering the plane without a scratch and for the achievement we had the privilege of experiencing. After flying 34 hours in some of the most beautiful country, we came home with the unanimous decision that if asked to do the trip again that the answer would be *yes*!

South for the Winter—Cross-Country Glider Tow

Ferry flights come in all sizes, shapes and even without engines. When the Connecticut Soaring Center of Plymouth decided

Fig. 8-13. Cub with floats on a dolly at Fairbanks' Phillips Field is an older version of the Super Cub/floats ferried from Minneapolis to Kenai.

they couldn't keep fooling around with Mother Nature in the winter, they sent a Super Cub tow plane and a Schweizer 2-33 two-place training glider to Florida for the winter. Riding the back seat of the glider was photographer John Harvey who penned his impressions for "Soaring Magazine." The enthusiasm of his glider flying experiences is infectious and well worth repeating.

Towline Adventure

Can you imagine:

—Spending 14 and three-quarter hours maintaining a precise high-tow position behind a Piper Super Cub towplane?

—Being like a sitting duck at the end of a 200-foot tow rope as a clutch of military jets decide it's time for a little mock target practice?

—Being towed just under redline a thousand feet above the Hudson River, over the George Washington Bridge, and out over New York harbor—hoping to God the towrope doesn't break?

Sounds like fun, huh?

All that and more were part of a crazy adventure I recently had the good (?) fortune to be mixed up in ferrying a Schweizer 2-33 the 1180 miles from a soaring site in Connecticut (Fig. 8-14) over 11 states to Florida.

It was two and a half days of boredom, tension and exhilaration—but mostly boredom, punctuated by more boredom.

Three of us went on the trip . In the Super Cub there was Art Cucuel, our ace towplane driver. In the 2-33 there was Geof Love, a jolly Englishman with over 2000 glider hours, and myself, Geof's student, whose longest previous flight had been the 58 minutes I did last fall to qualify for my B Badge.

Our decision to make this lengthy tow requires some explanation. There is a simple one—winter. Snow on the runway of our airfield, the Connecticut Soaring Center in Plymouth, CT, is sometimes so deep that the towplanes can scarcely get out of the hangar, let along launch gliders. And we have no plow.

While snow might keep sailplanes on the ground, it unfortunately doesn't keep the bill collectors at bay. So the people who run Connecticut Soaring decided that the time had come to follow Mother Nature's example and send some of their metal soaring birds south where they can be flown all winter.

They found a place called Indiantown, FL, 30 miles inland from West Palm Beach. It's in a part of Florida known for citrus groves, alligator-infested canals and—most importantly—600-fpm thermals!

Fig. 8-14. Super Cub and Schweizer 2-33 await takeoff from the snow-covered airport at Plymouth, CT (courtesy John Harvey, Figures 8-14 through 8-18).

The plan was to take some of the gliders down by trailer, but aerotow the rest—the towplanes had to go too, didn't they?

Although this wasn't to be the longest tow ever made, it was longer by far than any we had done. Who knew what problems might crop up? So it was decided to take a single glider down first, as sort of a dry run. If all went well, the rest would follow later.

The most critical part of the trip, we figured, would be getting past New York City (Fig. 8-15). In order to avoid a long detour around the New York Terminal Control Area, we planned on using the uncontrolled corridor that runs along the west side of Manhattan, 600 to 1,100 feet above the Hudson River. Since this locale doesn't exactly abound in the plowed fields that gliders favor for emergency landings, we sent a scout down in the towplane a few days prior to our departure to see if there was *any* place the glider could be put if the tow rope broke (Fig. 8-16).

It turned out that there were a few spots that might be landable if our short field technique was sharp—some piers, a highway, a railroad yard. Remember that gliders can be set down in a good bit less space than power planes can. After touching down at 45 mph or so, you quickly pull the spoiler handle fully out, activating the wheel brake, and if necessary push the stick forward to nose the plane on its skid. In a few seconds, and perhaps 100-200 feet, you are stopped.

Should a rope break occur at 1000 feet or so, your 23 to 1 glide ratio will carry you better than four miles and keep you in the air five minutes, giving you a little time to consider your options.

It was with reasonable confidence and a brand new nylon tow rope that we began our trip on a cold December day. Taking off at dawn from our snow-covered airfield, we headed south. Scarcely

Fig. 8-15. New York City in the haze, as seen by photographer John Harvey from the cockpit of a Schweizer 2-33 en route to Florida for the winter.

over an hour later, we had safely passed New York City. More on this later.

Reviewing my trip log, I see that our progress went as follows:

First day Plymouth, CT, to Easton, MD, fuel stop—3 hours 25 minutes; Easton to Knightsdale, NC, fuel stop—2 hours 59 minutes; Knightsdale to Dillon, SC, overnight stop—1 hour 15 minutes.

Second day Dillon to Richland, GA, fuel stop—2 hours 50 minutes: Richland to New Smyrna Beach, FL, overnight stop—3 hours 15 minutes.

Final leg New Smyrna Beach to Indiantown, FL,—1 hour 57 minutes.

The stops turned out to be interesting. On our very first one we walked into the pilot's lounge while the tug was refueling and Geof met a fellow member of the London Gliding Club—this in Maryland, USA, yet.

During another one, while making an unassisted (sans wing-runner) takeoff from a narrow runway, we almost knocked out a runway light, and our wingtip, before we had built up enough speed for the ailerons to level the wings.

After landing at New Smyrna on the second day, we were met by an airport official and chastized for breaking the airport rule against landing on the grass. Landing on the grass is standard procedure for gliders in order to minimize wear on nose skids and wingtips. But perhaps they weren't used to getting gliders at New Smyrna (Fig. 8-17). Anyway, the next morning we fixed 'em—we took off from the taxiway—who needs runways?

Aside from providing a welcome break from the towing, the fuel stops gave Geof and I a chance to shed layers of winter clothing as we got farther south. Since a glider is naturally unheated except for the sun shining in the canopy, we had to dress warmly at the beginning of the trip. We even wore electric socks. By Georgia we had gotten down to sweaters, but the electric socks stayed on until Florida (Fig. 8-18).

Our airspeeds averaged 80-85 mph, well above the usual aerotow speed of 60 mph. Since the Schweizer 2-33 cannot be trimmed for such a high speed, we rigged up a bungee (actually a screen door spring) that could be quickly attached to or detached from the control stick; otherwise the necessary forward stick pressures would have been unbearable on such a long tow.

As for ground speeds, our average over the 1,180 miles was 80 mph—not too bad. But on the second day, fighting a 15-mph headwind we watched trucks on Interstate 95 passing us. Our total progress for the day was a measly 450 miles.

Fig. 8-16. Down the middle of the Hudson River, as seen from the back pit of the Schweizer 2-33 sailplane. This is no place for a tow rope to break.

What's it like sitting for hour after hour on tow? Monotonous, mostly. We found that if you kept your eyes on the towplane too long you could experience "towplane hypnosis," which is sort of like "highway hypnosis." And even with the special bungee, our arms did get tired of fighting turbulence. If you contemplate making a long aerotow we suggest you instruct your tow pilot to fly above the convective layer, if possible. And for heaven's sake have him fly around any columns of smoke in your path; those "stubble-bubbles" can contain powerful, turbulent updrafts.

Luckily, there were two of us in the glider, so we could share stick-holding duties and also have someone to talk to. At one point my conversational talents must have flagged—Geof started kicking the rudder pedals for entertainment.

But don't get the idea that a long aerotow is always dull. There was that passage of New York. Imagine! There we were, a thousand feeet above the Hudson River, going right over the George Washington Bridge, which was jammed with morning

Fig. 8-17. New Smyrna Beach Airport with tow plane in the foreground during a ferry flight to Indiantown, FL, from Plymouth, CT.

Fig. 8-18. Destination airport for cross-country glider tow was Indiantown, FL. This was the conclusion of a six-hop ferry flight from the cold weather of Connecticut.

rush-hour traffic. Later we learned that we had been observed by a traffic helicopter and duly pointed out to the commuters below—what a sight we must have been.

And what a view we had of Manhattan, dramatically backlit by the early morning sun. We flew past the Empire State Building just below the level of its antenna; we passed the World Trade Center, which towered above us at 1465 feet; then we flew out over the harbor, going over Ellis Island, the Statue of Liberty and a Staten Island Ferry.

Then there was the time we were buzzed by the four jet fighters over Huggins, SC. It was our second day out and we were towing along as merrily as could be when two small specks appeared on the horizon. Soon there were two more specks. The specks grew and turned into A-7 Corsair jet fighters. We watched nervously as they sped toward us and flew a wide sweeping circle around us. Then it dawned on us: they were having a little unofficial target practice—and we were the target! Geof put his hand back on the release knob as they came back on their next pass. They split up as they approached, two passing right under us and two behind us. Then they flew off and we never saw them again.

And of course there was the time we passed the 15,000 feet space shuttle runway, longest glider runway in the world.

Fig. 8-19. BO-209 en route to Atlanta, GA. Nose gear retracts on this import while the main wheels are "down and welded."

Suffice to say that it was a memorable trip, well worth taking part in—once!

No Nosewheel in Sight: A Novel Import

There are many queer birds in aviation, but few of them have FAA commercial certification. Burt Rutan's homebuilt "VariEze" models are flying around quite comfortably with a fixed main gear and a retractable nose gear, but the only certified "store-built" model I know of with such a configuration is the import Messerschmitt-Bolkow-Blohm BO 209 Monsun. That's quite a mouthful, and so is the airplane (Fig. 8-19).

I first saw the new BO 209 at an AOPA Plantation Party in Las Vegas, NV. One thing led to another and it seemed that the dealer, located at the Georgia Peachtree Dekalb Airport just north of Atlanta, would kind of like to go home by airline. I offered to bring it back for expenses and promptly arranged to ferry a Cessna 172 back from Wichita to show a little profit on the week of funflying.

There's really not all that much fun in ferrying two airplanes across this country in a week alone. Therefore, I "conned" Rudy Adler, a fine aviation sheetmetal specialist with whom I'd worked at Fletcher Aviation into coming along. He could use the exposure and dual instruction to keep him current in his T-18 homebuilt.

When you read any aircraft sales brochure, you're exposed to superlatives that are outstanding and perhaps optimistic. The BO 209 brochure advises that: it is a fast all-metal, two-seat aircraft for sports and travel. The modern design details, such as retractable nose gear, incorporate the latest state-of-the-art improvements. Two reliable Lycoming engines are available for selection; a 150 or 160-hp engine (the latter with fuel injection).

"The tinted, full-vision hood slides back for easy access to the luggage compartment and to the two side-by-side seats. The electrically-operated landing flaps can be arrested in any intermittent position at any point throughout their range.

"The Monsun has foldable wings, eliminating the need for expensive hangaring. With wings folded, the Monsun requires but a fraction of the space occupied by other aircraft of similar size. Should the hangar be full, the Monsun can be towed on its own main gear by a car to be parked in the owner's garage. The wings are attached by three bolts and can be folded and unfolded easily in a few minutes. All electrical connections, fuel lines and control runs are coupled automatically. The entire folding mechanism is completely foolproof—fulfilling the most stringent certification requirements."

True, the BO 209 was a good little airplane, but it didn't turn out to be a real winner in the American sales market.

Adler and I flew N209HE in the Los Angeles area for a day or so just to explore its unusual handling characteristics and then launched off for the deep South. We spent our first night on the

Fig. 8-20. Rudy Adler, right, with the friendly FBO at Ranger, TX, prior to an early-morning takeoff for Georgia with the BO-209.

comfortable shores of Lake Havasu on the Colorado River and went on from there.

The import two-seater really did have good performance and range. We went non-stop to Cochise, AZ, at 150 mph on about 7 ½ gph. Then to El Paso, Sweetwater, and eventually Ranger, TX where the motel and barbecue shack were within walking distance of the minimal airport.

My notes show that Ranger had the brightest beacon in Texas that can be seen for 60 miles (Fig. 8-20). They also indicate that steak and eggs were $1.10 and ham with two eggs was 95 cents. Either this trip was a long time ago, Ranger was far off the freeway or gave a special discount to itinerant pilots. The FBO at Ranger told us that "no one comes to Ranger unless they're lost." There were no doors on the hangars and no reported thievery.

I've found that AOPA's annual *Airport Directory* is a useful adjunct to a ferry flight manual even though it is the size of a small phone book. This publication becomes a "wish book" when you head into the sunset (or vice versa) and try to decide just where to roost for the night. Prime considerations are courtesy motel pick-up, restaurant, good airport line service for fuel, and an FAA FSS nearby if at all possible. Ranger had a close-in motel with an adequate barbecue diner and lousy TV reception.

Next day Adler had the opportunity for good cross-country dual instruction where he had a chance to pick and choose the weather. He picked and we chose to wait in Cleburne, TX, for the

Fig. 8-21. Novel import aircraft always draw a crowd at the gas pit. The BO-209 is no exception during fueling at Ruston, AL. "Greenhouse canopy" is hot on the ground in summer weather. Most ferry pilots use a baseball cap or other head protection in flight.

Fig. 8-22. Rudy Adler and dealer Whitey Moore inspect the novel wing-folding mechanism of the two-place Messerschmitt BO-209 Monsun at the conclusion of the ferry flight at the Peachtree-DeKalb Airport, Atlanta.

fog to clear and then plugged along to Ruston, AL. At Anniston, AL, we landed to make a telephone call to get us into the tower-controlled airport with no radio and final delivery of the unique, import airplane (Fig. 8-21).

After an evening among the peach blossoms and Southern drawl—isn't it great the way people in different parts of the country speak the same language in such a different way—Adler and I boarded one of Delta Airlines' noisy Convair 990's for Kansas City and on to Wichita (Fig. 8-22).

At Cessna there was a 172, N4278Q, that was just Adler's size. I curled up in the right seat and dozed to Gage and Pampa where we enjoyed good Texas steaks. From Pampa, it was on to Tucumcari, Albuquerque and south to Truth or Consequences where hail and snow squalls cut the trip short after six hours and 15 minutes in the air that day. There was a note on the gas pit at Truth or Consequences that advised, "If unattended, serve yourself and leave a note with the number of gallons and type of fuel. I'll bill you." When we talked later with FBO Carl Hern, we were advised that "he had never been taken." Yes, those were great flying days!

Another five hours and 45 minutes later, plus fuel stops at Phoenix and Auga Dulce, CA and the new 172 was on delivery flight line in Long Beach.

As you fly more and more of these trips, in two-placers, prosaic, pedestrian production aircraft, or something more sophisticated, it would seem that there is an increasing feeling of professionalism each time you cross the Mississippi, see the first lumps of the Rockies and look again at the silver sliver of the Colorado River. For me and for many others who ferry airplanes, there is a feeling of sharing something extra special when you take another pilot (or pilot-to-be) along and expose them to this great picture of the face of our United States.

As pilot in command, you control the course, the vistas and the basic experiences (Fig. 8-23). Flight planning and route are dictated, as always, by the passing whims of Mother Nature who seems to assure that no two trips are ever the same.

Experimental on Cross-Country: This One Carried Guns

During the Korean Conflict, I was a test and demonstration pilot for Fletcher Aviation in Rosemead, CA. One of our projects was the FD-25B, a single-seat, close-in ground support airplane designed by John Thorp. Thorp later developed the popular T-18 homebuilt design that in many ways resembles the Fletcher fighter.

In the course of giving flight demonstrations to U.S. and NATO high-ranking officers, I ferried the FD-25B from coast-to-coast repeatedly (Fig. 8-24). On the first trip, Jerry Barden, Fletcher's project engineer and mechanic Bob Collins, with his 100 pounds of tools, accompanied me with the company Navion. Since the FD-25B had considerably better performance than the Navion, it was usually Barden's assignment to do the navigation and I would tag along off his wing at varying distances. Just to make the headwinds across Texas less annoying, I would "park" off the Navion's wing and pretend to read a pocket book just to keep Barden and Collins on their toes. Since the "Experimental" FD-25B required a parachute, I was taking advantage of a situation, but in reality I was cautious about the whole thing (Fig. 8-25).

However, most of the cross-country trips in this armed single-seater were made alone, without benefit of escort. At the time I was flying it, the FD-25B actually had two 30-caliber machine guns on loan from the U.S. Navy mounted in the wings, designed to fire outboard of the prop arc. During one long tour in Washington, D.C., I was flying regularly out of the Washington National Airport. One day President Eisenhower came to the field to bid goodbye to his brother who was going out on the "Columbine," then Air Force One. I figured that some of the visiting

Fig. 8-23. The brilliant red spires of Arizona's Monument Valley are never seen better than from a light airplane. This view is always inspiring as pilots ferry light aircraft to the West Coast.

military brass might be interested in looking at the ship, so I called ground control and received permission to taxi down near the presidential plane. As I spun the taildragger around, it occurred to me that the security forces would have had a complete panic had they known that the ship was armed.

We had the 25B painted jet black at the time (Fig. 8-26). One of our sales pitches for the armed light plane was to counteract the

Fig. 8-24. Fletcher FD-25B ground-support airplane flown by designer John Thorp. Note design resemblance to Thorp's popular T-18. Power was a 225-hp, six-cylinder Continental engine.

Fig. 8-25. Fletcher "jet-cooled" Navion used as a support airplane on demonstration flights. (Photo was taken from the Fletcher FD-25B.) The augmeter cooling system used on the Navion and sold to some owners was similar in design to cooling of the FD-25B.

North Korean's "Bed Check Charlie," one or a series of old biplanes that cruised over the lines at night and threw a few hand grenades out of the cockpit. There was little damage and no reported casualties, but the operation did cause some confusion.

While we couldn't sell the concept to this country, one FD-25B was eventually crated and exported to Cambodia where we heard five units were built. Fletcher's engineering department

Fig. 8-26. FD-25B with its black paint job would have been useful in Korea to combat the nuisance raids of "Bed Check Charlie." The author flew for Fletcher Aviation for 2½ years, much of it demonstrating this airplane to potential military customers.

Fig. 8-27. U.S. Army high-ranking officers inspect the Fletcher armed light plane at Old Fort Comfort, VA.

Fig. 8-28. Instrument panel of FD-25 during a climb past 5,000 feet. Fun sight in the center of the panel was for two 30-caliber machine guns.

had subsequent correspondence asking how to repair small arms fire damage, but we had no details on the tactical use of this little fighter (Fig. 8-27).

On one cross-country trip, I had an appointment at the Orliken Rocket Factory near the Black Mountain Airport in the hills of Kentucky—well in the moonshine-making country. At the time, the ship was carrying two 40-gallon napalm drop tanks. The tanks were empty and I made every effort to purchase 80 gallons of good corn "likker" to ferry home. It would have made one hell of a company picnic. Fortunately, or otherwise, I couldn't find a moonshiner and came home empty (Fig. 8-28).

On one of these trips, I was headed back toward the Los Angeles area after a long day in the air. I was at 10,500 feet just west of Blythe, CA, when a huge black column of smoke blossomed along the highway. From my altitude, I wasn't sure what had taken place, but I attempted to advise Blythe Radio that some sort of an explosion had occurred. The single Mark 12 radio wasn't doing a good job and I wasn't sure that the message had been received.

In the local paper the next day, I read of a head-on crash along the highway involving a fuel truck and trailer. There were a number of fatalities.

It was perhaps two months later when Fletcher's lawyer called me into his office. "By the way," he said, "did you happen to report a gasoline truck crash on your last ferry flight from the East?" I nodded and looked puzzled, particularly because of the time delay.

"I just had a lawyer call me. He's representing the family of one of the victims. He'd checked back with the FAA on the original report of the crash and all that they had was a garbled message with the partial call sign 'one-six.' The ship was N91316. It took the FAA this long to track down just which airplane had made the report. The lawyer wants to know if you can make an eye-witness statement about the crash."

I assured our lawyer that there was no way I could have seen anything before the crash from 10,000 feet above the desert.

So you never know what sort of a follow-up you'll have from a routine radio call on a routine ferry flight.

Chapter 9

Wings Over Water

The increased dependability of today's engines and airframes makes long over-water ferry flights relatively routine. A detailed report of one such commercial ferry service, prepared for Plane & Pilot magazine, was one of the developments that caused us to write this book at this time. No, we didn't offer to fly such a delivery, but we were honestly tempted.

Passports, visas and those ever-lovin' shot records apply to pilots on almost all over-water flights. They're not included here since this is required by all professional ferry pilots. You'll find details of international paper work recounted in the next chapter on a Cessna 207 that we delivered to Chile.

Any pilot who gets very involved in ferrying should, of course, keep his or her passport valid and keep up-to-date on a few of the basic shots.

We've touched on over-water ferrying to Australia with a U.S. Army C-47 in an earlier chapter. Unless you've flown that broad Pacific in something smaller and slower than today's jet-age shuttle, you really can't appreciate just how big that ocean really is.

By contrast, the North Atlantic trip is a series of fairly short hops as detailed in the Turbo-DC-3 report earlier in this book. Even the middle Atlantic run, made once in a Tiger DC-4 from Bermuda to the Azores, Lisbon and Rome and back to Caracas, Venezuela, really isn't all that long a flight. It was only 8:10 from Bermuda to Caracas in the DC-4.

Therefore, the 21- to 24-hour flight from Santa Barbara to Honolulu, first leg of a light plane delivery to Australia, is really a big hunk of the world to tackle on one hop in a small airplane. Meet the ferry service that regularly flies the longest over-water hop in the world, from California to Hawaii, and then to Australia.

How Southern Cross Aviation Does It

The longest overwater ferry flight in the world begins on the Pacific Coast of the U.S. and heads for the South Pacific. It begins at approximately 6 p.m. and the first leg is 22 hours. Ferry flights with single-engine aircraft to the South Seas are becoming relatively routine—that is, if you consider the very wet 2170 nautical miles between Santa Barbara, CA and Honolulu as a routine single-engine trip.

Southern Cross Aviation, operating out of a single hangar at Santa Paula, CA, averages eight to 10 deliveries a month to Australia (Fig. 9-1). They have 10 "permanent pilots" who contract to make the overwater trip and send the ships out in groups of three or more. One recent flight was of 10 aircraft, broken into two flights of five each.

There's safety in numbers and the over-water aircraft fly formation on the entire trip. Most flights are planned at night for better weather, better radio reception and more airliners overhead. Out of the first 350 single-engine aircraft ferried by Southern Cross, only one was ditched.

South Pacific deliveries from Santa Paula take up to 75 flying hours and seven days' travel. Each formation must be keyed to performance of the slowest aircraft in the gaggle. On the flight of three shown in the illustrations, the Piper Warrior was the slowest of the group with the Cessna 182 and Piper Lance appreciably faster. All aircraft were equipped with ADFs and at least one aircraft in each group carries a Collins HF 200 portable HF transceiver. This redundancy in equipment proved most helpful when the complete electrical system on three out of four single-engine aircraft quit on the inbound leg to Honolulu, but the formation continued without incident. Later, it was discovered that one worker at the factory—unnamed, since the problem has been corrected—was installing the main electrical lead-in without sufficient slack in the cable. Given the vibration of some 20 hours of flying, the cable broke.

Southern Cross is headed by two young transplanted Australians: Peter Dickens and Graham Butler (Fig. 9-2). Their opera-

Fig. 9-1. Three Southern Cross light planes head out over the Pacific Ocean for Australia. The next land they will see is Hawaii.

tion is unique in many ways. Not only do they handle the ferrying in many cases, but they also purchase the aircraft in the United States and resell them in Australia. Almost all aircraft are new with only ferry time from the factory to California. Some twins are purchased used; but if the engines have more than 500 to 800 hours on them, they are replaced or zero-timed before leaving California.

"We can save 25 percent by installing new engines here," explained Peter Dickens, who carried the title of Vice President, Operations. "Since engine components are readily available here, we can replace the units and provide the customer with a brand new powerplant package. It's also safer for our pilots."

Fig. 9-2. Peter Dickens, of Southern Cross Aviation, and his dog run errands at the Santa Paula Airport. The Skylane and twin Cessna are headed for Australia. The author's Cessna 170B (at left) is not.

Southern Cross moved from the busy, expensive Burbank Airport to the private field at Santa Paula, CA because of economics. The airport is large enough to handle both single-engines and twins, but hangar rent is 50% less. Also, "There's a lot of talent right on this airport," said Dickens. Four small businesses, all in the same hangar row, provide essential services to Southern Cross.

Flight Communication handles the specialized radio installations in the hangar space immediately adjoining Southern Cross.

Jim Kern and his son Ric, of Profab, have developed baffled galvanized-iron tanks for these deliveries. The tanks are strictly a one-way operation since they cannot be returned after having been filled with fuel. The Kerns build an average of 30 tanks per month in sizes varying from 50 to 130 gallons. A fixed-gear Cessna 182, for example, will carry two tanks in the cabin totaling 180 gallons of fuel.

Larry Endicott, of Crystal Air, works in a nearby hangar and installs all the tanks after they have been pressure-tested to 3.5 psi. Tanks come in up to 10 basic sizes and shapes. Plumbing differs for almost every make and model (Fig. 9-3). Braided steel cables wrap around each tank installation and an FAA inspection is required for each new type of tanking. Larry Endicott and Jim Kern took predictable pride in the tanking of the Cessna 185 that was ditched. Even though there were 110 gallons in the temporary fuselage tank when the ship hit the water, the tank didn't burst, and the tiedowns did not break. The ship did sink within 15 minutes, despite a hawser around the tail from the rescuing freighter, but the pilot was in the water only three minutes.

The meager statistics available on ditching fixed-gear airplanes indicate that a high-wing aircraft, ditched properly at minimum speed, will not flip on its back because of the large area of the wing that impacts the water after the landing gear is either bent or sheared off by initial water contact.

The ferry tanks from Santa Paula are given away in Australia. They have proven to make excellent storage cabinets.

Dickens, a former Australian Air Force pilot and member of the exclusive Society of Experimental Test Pilots, readily concedes that, "It's safer to go across the Pacific in a single-engine airplane than a twin. We have an FAA waiver for up to 25% over gross on a ferry flight. It is doubtful if a light twin can fly on one engine just after takeoff with the amount of fuel required for the Honolulu leg. We presume that a twin Cessna Skymaster, for

Fig. 9-3. Peter Dickens crawls out over a temporary fuselage tank in a Cessna 402. In case of ditching, pilots are advised to smash the windshield with an ax and exit from that end.

example, will not stay in the air with the 500 gallons of fuel required until 100 gallons have been burned off. Therefore, you're flying for the first three to four hours with less safety than with a single engine since you have two that can quit." (See Fig. 9-4).

Dickens learned celestial navigation in the Australian Air Force and has one other pilot, Keith Rasmussen (an ex-Navy Reserve P3 pilot), who also can navigate by the stars. Southern Cross reports that several of their pilots are working with the sextant between flights and hope to become proficient in this proven type of long-range navigation.

During the Vietnam War, hundreds of Cessna 337s (OV-2), the military version of the retractable-geared "push-pull," were ferried to the combat zone. On those ferry flights, pilots were paid all expenses and about $1,000 per trip. Engineering and subsequent in-flight analysis indicated that the Cessna OV-2 would not fly on one engine at the over-gross takeoff weight of 5,800 pounds until at least 100 gallons of fuel were burned off. Therefore, for the first seven hours of the Honolulu leg, the ship was going into the water if either engine failed.

Southern Cross considers paying ferry pilots a lump sum for a delivery as a safety hazard since pilots will tend to overfly airports where fuel is expensive and to "push" to keep the number of days to a minimum. The ferry service pays the fuel, furnishes all overwater survival equipment to assure that it is current and operating properly, picks up the airline fare back and pays the pilot a regular

Fig. 9-4. A Cessna Skylane II complete with ferry tanks and over-water survival equipment is readied for takeoff by Southern Cross.

fee. Dickens and Butler won't divulge the actual payment for each flight. The ferry service covers all fuel costs and delays not caused by the pilots. Mechanical and weather delays are covered. Shopping, a ferry pilot's Whim, or other entertainment is charged to the crew. It's all a fairly GI operation.

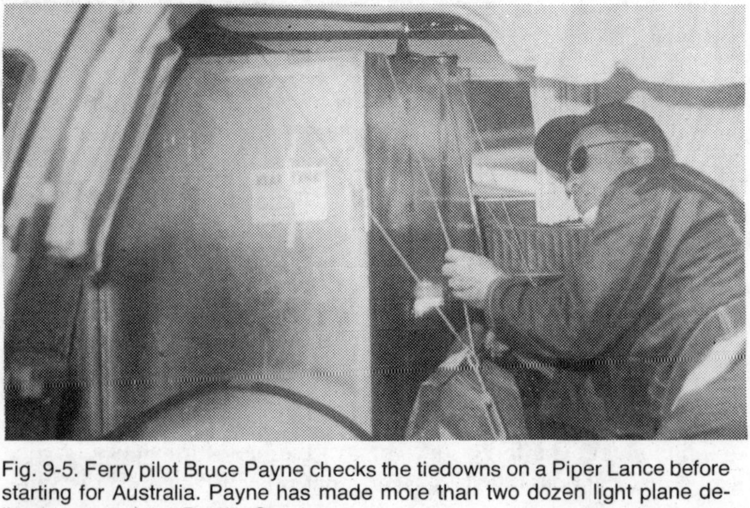

Fig. 9-5. Ferry pilot Bruce Payne checks the tiedowns on a Piper Lance before starting for Australia. Payne has made more than two dozen light plane deliveries across the Pacific Ocean.

We follow a flight of three. Lead pilot in the Piper Lance is Bruce Payne, an ex-U.S. Air Force pilot with 22,000 hours. Bill Friel, flying the 182, is ex-U.S. Navy with more than 5000 hours, while Keith Finley, Army Reserve pilot, has 3500 hours (Fig. 9-5 and Fig. 9-6). Southern Cross wants at least 2000 hours flying time with a reasonable amount of actual IFR.

Southern Cross rents its hangar from Mira Slovak who flies for Continental Airlines. As the trio checked their three aircraft and loaded the last of their supplies, Slovak stopped in with the airline version of the latest weather. He would be flying to Hawaii this same evening and would be talking with the crews en route. The unofficial overwater frequency in the Pacific is 123.45, but call-ups can be initiated on 121.5 with a change-over to some other frequency. The ferry pilots will talk to as many as 20 air carriers during their long hop to Hawaii. Navigation at night is made easier by the strobe lights of the airlanes and by their controls in daylight. Because of the cloud deck frequently found over the Pacific, it is not unusual for a ferry pilot never to see the water from after takeoff until reaching the Islands.

The Collins HF unit, which is carried in at least one plane on the flights to Australia, is standard on all twins (Fig. 9-7). There is a punch-card frequency selection on the solid-state equipment, and Southern Cross programs at least one Amateur Radio (ham) frequency into their ferry flight frequency card. On a recent flight, one VHF radio went out midway to Honolulu. The flight leader

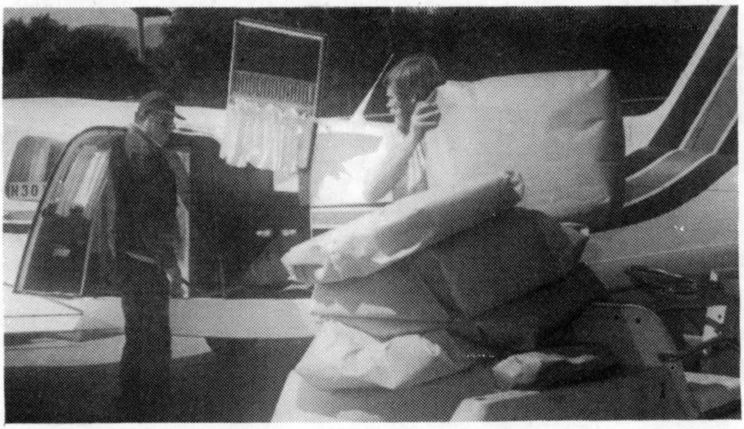

Fig. 9-6. Ferry pilot Bruce Payne watches as Peter Dickens of Southern Cross Aviation loads seats into what space remains in the Piper Lance headed for Australia.

called a local ham operator in Santa Paula who put in a phone patch and woke up Peter Dickens in the middle of the night. The Operations Director was able to talk directly with his pilot near the "Daisy" checkpoint and find out what radio was malfunctioning. A replacement unit was hand-carried to the Los Angeles International Airport where a commercial jet was headed for the Pacific. A crewmember took the unit and delivered it in Honolulu to the waiting ferry pilot at 9 p.m. for a 10 p.m. departure. Without the HF radio and ham operator hookup, the flight would have been delayed at least a full day awaiting replacement parts.

An important flight-planning and en route aide to the Southern Cross pilots is the latest available infra-red satellite photo of the Pacific Ocean (Fig. 9-8). These are supplied on a regular basis direct from the Honolulu NOAA station. The satellite photos, similar to sequence coverage used by many regional television stations on their weather shows, enablc pilots to plan their route, even if the NOAA prints are 10 to 12 hours old. On the infra-red photos, the darker the color, the warmer the temperature. Therefore, cloud build-ups with towering tops can be evaluated directly from the photograph.

After the ships are loaded—usually every nook and cranny in the cockpit is filled with spare parts, seats, life raft, charts, personal survival gear, a single suitcase and at least one empty gallon plastic water jug for use as a "range extender"—and after a final briefing on radio frequencies and formation assignments, the three ships taxi the length of Santa Paula's 2800-foot strip and are

Fig. 9-7. Collins HF-200, installed permanently in a Cessna 402, is programmed before the ferry flight for the various frequencies to be used. At least one amateur radio frequency is on the list.

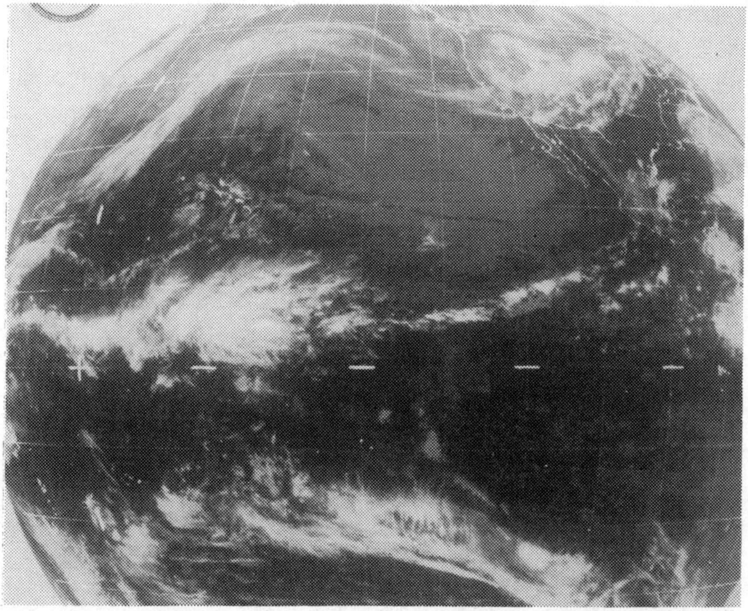

Fig. 9-8. Infra-red satellite weather photos are used for ferry flight planning. This NOAA photo shows a normal summer intertropic cloud pattern over the Pacific. A fairly accurate measurement of cloud heights can be obtained from the degrees of gray in the clouds. The warmer the temperature, either in cloud or water, the darker the area in this black-and-white photo (courtesy NOAA Satellite Service in Honolulu).

soon airborne. All run-ups are made while moving to eliminate prop damage.

It's once around the field and a fly-by in formation at pattern altitude (Fig. 9-9). Santa Paula is a small town with an enthusiastic group of general aviation pilots who base at this privately-owned airport. They have become accustomed to this departure fly-by. Then it's up the coastline 35 miles to Santa Barbara's Goleta Airport. The FAA FSS there has a special briefing transmitted from San Francisco where the most detailed trans-Pacific weather data is gathered. All tanks are topped and then it's time to go.

Low-level weather reports sometimes leave a little to be desired. One flight of five 172's and 182's were expected to take 20 hours to reach Honolulu. There was a wind shift and they took 25 hours, but still had three hours of fuel remaining. Pilots run a careful fuel consumption check at their 6000-foot cruising altitude at the power setting they plan to use and carefully measure the amount of fuel that goes into that one test tank. Their flight is planned to use fuel from the temporary ferry tanks first and save the regular tanks for reserve. No additional oil is carried and no provisions are required to add oil in flight since all the engines are new except for ferry-break-in time.

Southern Cross pilots rapidly develop more night formation flying, both in and out of clouds, than most military pilots ever log. Use of strobe lights at night is the option of the individual pilots and the flight commander. Each group has a flight leader, structured much like a military operation. The leader calls all the shots and the other pilots follow instructions. After three or four flights, pilots are taken on a delivery, usually with Peter Dickens as lead pilot under supervision. Southern Cross attempts to keep five lead pilots qualified (Fig. 9-10).

From Honolulu, the aircraft normally bypass Christmas Island and a "sometimes" ADF homer, both because fuel costs $3 per gallon and more importantly because tailwinds of up to 20 knots can be expected at the 6000-foot cruising altitude from Honolulu to Majuro in the Marshall Islands—18 hours at 100 knots. The next stop is 11 hours to Henderson Field at Guadalcanal. From here the aircraft either fuel at Port Morsby, New Guinea, or go directly to Brisbane, Australia. The overland leg to Melbourne is another eight hours where Schutt Aircraft Sales takes most of the new aircraft.

What It Takes

Who really wants to be a ferry pilot on such an adventure in a single-engine aircraft? One national magazine report titled "The Australian Connection" produced more than 500 letters and an equal number ? of phone calls asked for a flight. "One girl named Bunny advised us in writing that 'she'd do anything to make one of these trips,'" grinned Dickens. "We didn't hire her. Many of the pilots had good qualifications, but only wanted to make one or perhaps two trips. We finally answered the phone calls by asking how much flight time the pilot had. If he said 1500 hours, we told him we needed 2000. That worked fine until a retired airline captain said he had 30,000 plus."

"Actually," Dickens continued, "We want regulars who are accustomed to our rather rigid type of flying." Bruce Payne, for example, who headed the three-plane flight shown in these pictures, is a 22,000-hour military lieutenant colonel who flew KGIL's "traffic watch" for seven and a half years before Francis Garry Powers was hired as a back-up pilot. Payne has made 26 trips to Australia in about two years for Southern Cross.

Fig. 9-9. Three Southern Cross Aviation ferry pilots head over the orange groves near Santa Paula, CA, en route to Australia. Flight leader in the Piper Lance is Bruce Payne. Bill Frier is flying the Skylane and Keith Finley is in the Piper Warrior.

Payne explained that the overwater night formation flying was relatively simple with a full moon, but could be fatiguing on a black night. Pilots turn on the under-wing "courtesy lights" on the new high-winged Cessnas. When the formations encounter clouds, Payne has each wing man turn away two to three degrees for at least five minutes and drop or pick up 300 to 500 feet. Then as the ships break clear of clouds, they reform.

Normally, pilots do not attempt close-formation letdowns, but Dickens told of one flight of seven 172's that made an approach together into Nandi on Fiji. The Nandi airport has "bad terrain" with 5000- to 6000-foot mountains on one side and the sea on the other. In the mass formation let-down into Nandi, Peter Dickens was flight leader. The aircraft were positioned in a double-diamond formation with the pilot who had never been into Nandi before in the tail-end-Charlie position. Ed Brunner, owner of Centair Aviation, Memphis, TN was making his first trip and for some reason didn't have a let-down chart for Nandi. When Brunner radioed to Dickens, "What happens if I lose contact?" the flight leader's reply was succinct: "Just don't lose contact!"

Southern Cross ferries a number of Maules to Australia where some of the aircraft are used for glider towing. Peter Dickens is thinking about towing a sailplane on one of these South Pacific flights. The main obstacle would be in positioning the sailplane properly on tow in instrument conditions. To compensate for a slower tow speed, additional fuel would have to be carried in the sailplane and transferred by flex line to the "Tug." There are a number of problems to be resolved before this "stunt" can be attempted.

Southern Cross finds that their work is somewhat seasonal. Their busy season is April through October when weather all along the route is better. Some flights have had to wait as much as four weeks to depart Santa Barbara with an acceptable wind component. This busy season also coincides with the Australian fiscal business year that ends on June 30th. Since the timing on many aircraft purchases are based around tax billing, Southern Cross expects 75% of its deliveries during the summer season.

The complete cost of the delivered airplane is "comparable" with surface shipping, but savings in time and reassembly make the ferry time on the aircraft more than worth the trip. "Since most all the new aircraft are painted with polyurethane that lasts almost the life of the aircraft, it is important to the buyer not to have

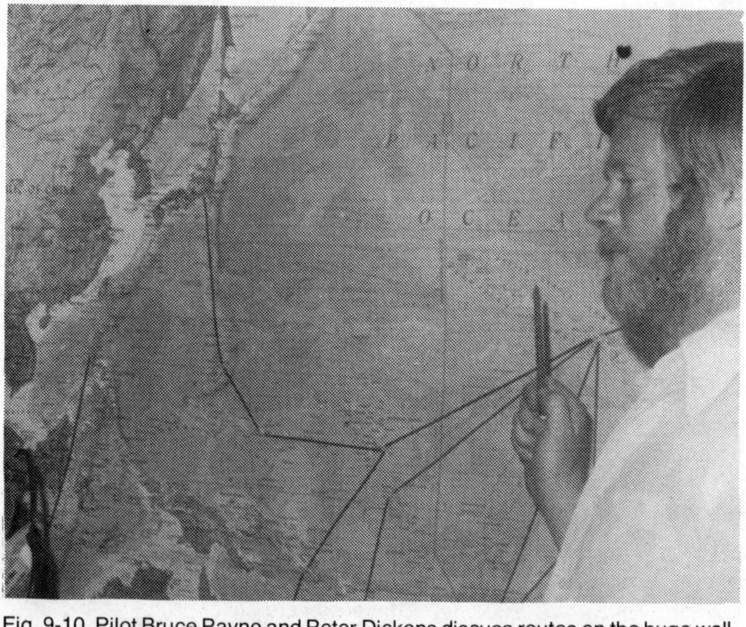

Fig. 9-10. Pilot Bruce Payne and Peter Dickens discuss routes on the huge wall chart before a ferry flight to Australia. Fuel cost and weather are prime factors in routing.

Fig. 9-11. Southern Cross pilot Keith Finley in the Piper Warrior flies off the wing of flight leader Bruce Payne. After topping all tanks in Santa Barbara, next stop is Honolulu.

screw holes show and have the scratches that invariably come in reassembly," explained Dickens. "Of course, the time delay is important, particularly when the aircraft in a crate isn't making any money for the owner. Then there's that assurance that we've taken the best possible care of the new plane. Otherwise, we wouldn't be ferrying it over all that water. There is no depreciation in Australia for the time that is logged in ferrying."

Now that the route to Australia has been well-proven, Southern Cross is beginning to deliver airplanes to Japan via Honolulu (Fig. 9-11), Majuro, Truk and Saipan. They hope to ferry Australian twin-turboprop Nomad Aircraft back to the U.S. and come up with a lucrative two-way traffic. Money isn't made by sleeping in the coach section of a jet.

Deliveries are also planned to Europe with Islanders scheduled to return to Anchorage. Therefore, the ferry pilots of Southern Cross might not always have that well-tanned skin from repetitive RON's at Waikiki, Pago Pago and beautiful downtown Guadalcanal. It's all ferry interesting!

Chapter 10

Have Passports And Shots—Will Travel

Most international ferry flights are handled by established companies, either the small, one-man or one-woman organization or the larger companies that have become well-established in the business. There is a good reason for using companies who "have been there before." As the learning curve is flattened out, crews working for companies who regularly take aircraft to foreign destinations are able to complete flights on schedule and with a minimum of expense.

Any international flight requires a mass of pre-planning and paper work. You must check current health requirements for the part of the world you plan to visit, and these can change by the week. That yellow international innoculation record goes right with your passport.

On a recent out-of-the country ferry flight, we knew our smallpox shots needed updating and went to our personal doctor for the innoculation before we checked with the health department. We discovered belatedly that due to a yellow fever epidemic in two countries where we would be landing that shots would be required. The smallpox and yellow fever shots should have been given simultaneously and since they were not, there was a four-week wait from the date of the smallpox vaccination—and we almost didn't have enough time. Only the health department can give the yellow fever shots and it takes an appointment, at least in this part of the country.

And, of course, your passport must be valid and have the required visas. Many countries now require only tourist cards, but you really can't go far without a passport and it is a required identification for almost all the paper work you sign.

If you were merely planning a junket via airline, you could pass almost all this paper work along to the international air carrier. That's part of their ticket charge. However, most foreign countries with the exception of Canada and possibly Mexico, just aren't geared up to handle light airplanes.

Overflight authrization or advanced notice of landing is foreign—and foreign to most U.S. pilots. However, it exists and the established ferry company is familiar with current procedures. AOPA's Flight Planning Department is geared to handle this paper work both for transient pilots and on delivery flights. Anyone making an out-of-the-country delivery should take advantage of their highly professional services. It's worth the membership.

The latest in charts is also available in a package from AOPA. They can supply low-altitude en route charts while Jeppesen has the IFR "plates" in any area you request. As in a recent advertisement, "don't leave home without them!" Plan ahead and order all your charts well in advance of your trip because these charts are not always stocked in volume.

Survival equipment depends on your route and the time of the year. Over-water ferry or extended flight over jungle has a certain risk, and you should carry whatever you feel necessary to stay alive if you are forced down. Here again, the ferry service that has been there before either has the survival package or knows what it should contain and where to get it.

It is appalling how naive an experienced pilot can be about international flight until he's done it once or twice. The following report of a Cessna 207 (Fig. 10-1) delivery illustrates what little detailed flight information is available off the beaten path. Take the simple matter of visas and tourist cards. Rather than spend days hand-carrying our passports through the various embassies and consulates, Julia and I contracted with the Visa Service in San Francisco to handle this portion of the paper work. They advised that they were unable to obtain a tourist card for us for Colombia since we were not coming and going on a scheduled airliner. It took a phone call and subsequently a personal visit by both of us to the Los Angeles office of the Colombian Consulate, complete with passport and two photos, to obtain a special tourist card.

If you've never flown to South America before, you might go into the project just as fat, dumb and happy as we did. We scaled

Fig. 10-1. New Cessna 207 in flight. This ship is similar to the model ferried to Santiago, Chile, by the authors (courtesy Cessna Aircraft).

out the distances, checked them against the range and performance charts on the 207, and planned the trip just as we might here in the United States. There was littlc worry about "paper work" delays, misty weather forecasting, ample 100-octane fuel availability or any problems in charting that might exist in pointing out which are military airports and which are civilian, which "international" airports don't turn on their VORs or landing lights unless you call them on an obscure frequency and the airport manager isn't out for a cup of coffee.

AOPA's Flight Planning calls for both routing and itinerary so the appropriate telegrams can be sent for notification of over-flight or landing permission can be sent. This seemed straight forward enough, so we unrolled the charts on the living room floor, checked the various airports of entry against the range of the new Cessna, picked what looked like appropriate stops with enough time for sightseeing and submitted the list. Of the 16 airports listed in the original planning, we finally arrived at only seven of them, and not once on the day originally planned.

AOPA's excellent *Latin American Flight Guide* is a must for a southeast-bound flight. When you look at the charts, note that this trip goes way east with Santiago daylight saving time three hours east of New York. The *Latin American Flight Guide* doesn't continue beyond Panama because of the limited amount of private air travel in that area, but it is excellent as far as it goes. If you're headed in that direction, read it carefully.

Read as much as you can in advance about the countries you plan to visit. While there is little time for formal sightseeing on a ferry flight unless you're boxed in by weather or a mechanical problem—or in some foreign countries by the vagaries of fiesta, siesta, snarled paper work, etc.—you'll know more about what you're flying over with extensive preflight reading.

If you have foreign-made cameras, binoculars, expensive recorders or other import items, be sure to obtain and carry a U.S. Customs declaration form proving that these items have been imported legally and any required duty has been paid already. Otherwise, you can be required to pay that duty a second time. Frequently, a copy of your personal property insurance form with the items listed will suffice, but that depends on the individual inspector.

One ploy used by many pilots on overseas flights, whether on a delivery or just a visit, is to wear a uniform, complete with epaulets, wings, gold braid on the cap, and always call yourself "Captain." It works wonders in streamlining the mountains of paper work (Fig. 10-2).

One ferry crew we met in Wichita recommended a dozen copies of "Playboy" or any other such publication as an ever-ready means of greasing the paper mill. Experience dictates that these publications are often a more valuable give-away than cash.

Factory baseball caps or jackets are also much desired in out-of-the-way places and should be considered as a means of making complex situations less aggravating.

On the specific Cessna 207 delivery described here, the flight was made on a flat fee basis (Fig. 10-3). By the time we paid some whopping landing fees—$40 in Colombia, $56 in Ecuador and a fine of $75 in Peru—plus fuel that came to over $3.00 per gallon, we didn't cover expenses. Sure, the "experience" was interesting and informative, but it was also expensive.

There is still adventure, frustration, potential hazard and the possibility of no profit in ferry flying. Overseas and foreign flights present a whole new dimension to delivering airplanes that is best handled by the "pro's." In final research for this book, Julia and I contracted to ferry a new Cessna 207 from Wichita to Santiago, Chile. Come share our misadventures and we believe you'll think twice before volunteering for a similar ferry flight unless you are working for a ferry service that handles these assignments routinely. A condensed version of this report was published in Plane & Pilot Magazine.

Ferrying a 207 to Santiago, Chile

It was a nice time of the year for a comfortable, scenic ferry flight, so we arranged to deliver a brand new Cessna 207 to South

Fig. 10-2. "Tony" LeVier, Lockheed's famed test pilot, wears his Captain's uniform as he embarks on an international demonstration trip with one of the early 1011 tri-jets. The military-type uniform and name of "Captain" are useful at almost all foreign airports.

America. The new airplane was fine, but the fun and games we anticipated far south of several borders didn't turn out a bit that way.

If you plan to fly yourself south by southeast, take liberal amounts of money, patience and guts. Gather all the flying experience you can muster and hope for a fair proficiency in Spanish—or we wouldn't recommend this trip for you.

The new ship was headed for Aeroservicios Ltd., in Santiago, Chile (Fig. 10-4). There was a 97-pound box of spare parts stuffed in the back of the cavernous cabin. Up front, equipment included a single nav/com radio, an ADF, single-axis auto pilot, and long-range tanks with 73 gallons usable. Not included were a tur-

Fig. 10-3. Cessna 207, ferried by the authors, as it sits on the delivery ramp (front center) at Cessna in Wichita. Snow banks are just melting in Kansas, but it is summer in South America.

Fig. 10-4. Cessna factory photos of a "full-house" 207 and the cabin interior. Note that the aircraft delivered by the authors had considerably less equipment on the instruments.

bocharger, transponder, second VHF radio and glideslope, single-sideband or HF transceiver or handy-dandy, thumb-activated trim tab. Zero One Hotel had everything that was necessary, but not much more.

Pre-flight planning included AOPA's Travel Department for entry and overflight permits to the countries south of Mexico. This maze of paper work is awesome for a first-timer, but Catherine Howser, who has headed AOPA's department for many years, is up-to-date on the ever-changing requirements along this route. It was well worth every penny to have the majority of this detail work taken care of (Fig. 10-5).

The latest available charts also came from AOPA's Jake Goodrich. Some of these latest charts still carry a nine-year-old date, but there just isn't that much VFR flying in that part of the world. Essential are DoD low-altitude en route charts, a DoD Flight Information Supplement (FLIP), and a package of DoD or other low-altitude instrument approach "plates." You find the difficulty of not being able to get all the information you need from any one package. VFR charts have no nav-aid frequencies nor detailed airport information. Nose-to-nose pre-flight briefing suffers greatly with a language barrier. English might be considered the International Language of aviation, but more often than not it just wasn't available on the ground or in flight.

If there's more grey in the hair and more perspiration in the palms, it was not without justifiable reason. The saga of Zero One Hotel is certainly one of the unplanned misadventures in aviation.

In the front pit we had two licensed pilots with a total of 45 years of experience, 10,000 hours plus, all the ratings but ATR. Unsuspecting passengers as far as Lima, Peru, were relatives, Paul and June Crawford, from Santa Barbara, CA. They are a retired 70-years-of-age "going-on-40," and both have a fair to good knowledge of Spanish.

The actual trip began in Wichita with a routine checkride with Cessna pilot, Jerry Robinson, himself a veteran of several years of flying in South America (Fig. 10-6). He was sure that our shortfield landing techniques were up-to-snuff before giving us the green light.

After signing for the ship and picking up the appropriate papers, we went out to depart, but it took three radios to get one

JAN 24 1979
AOPA FLIGHT DEPT.
DP. DATE ________

ZCZC IWK251
FF MCBOYADO
241950 KDCAXHAO
PLS CONFIRM LANDING PERMIT FOR FERRY DELIVERY FLIGHT CESSNA
207 ACFT REGISTRATION N6401H PILOT DON DOWNIE PLUS ONE CREW
AND TWO PASSENGERS ALL USA CITIZENS

FEBRUARY 13TH FROM PANAMA CITY/LAND MEDELLIN/LAND QUITO STP

NNNN

✓COLOMBIA

JAN 25 1979
AOPA FLIGHT DEPT.
DP. DATE ________

ZCZC VIK356 252101
JJ KDCAXHAO
251612 MCBOYADO
AUT ATERRIZAJE EN MEDELLIN AERONAVE CESSNA 207 MATRICULA
N6401H FEBRERO 13/79 PANAMA/MEDELLIN/QUITO PUNTO
BL2048

Fig. 10-5. Copy of AOPA Flight Department request for landing at Medellin and reply in Spanish from Colombia approving arrival on a specific date.

that would really work. However, the third unit worked perfectly for the entire trip. We would have been in even deeper trouble at several places if the nav/com had quit.

Continental's ever-loving fuel-injected I0-520-F poured out its 300 horses on every leg. The deeper we went into the jungle or offshore to evade newly-created prohibited areas, the more love we had for that smooth-running powerplant. In fact, had we encountered any maintenance problems, either in airplane or equipment this trip would have been a complete fiasco.

So, let's get off the plains of Kansas, enjoy Cessna's legal buzz job at 300 feet for five miles to stay below the adjoining McConnell Air Force Base traffic, and head south by east. The first stop was to check fuel consumption, look for any leaks and get a late lunch was Ardmore, OK. Then on to Dallas-Ft. Worth International to pick up Paul and June. Our export 207 had no need for a transponder in Chile, so we talked to DFW approach, landed at Denton, TX, for a telephone briefing, and were cleared into the TCA without benefit of transponder.

The next morning, the four of us loaded aboard in a drizzly rain and waited until the ATIS raised a 900-foot ceiling to 1100 feet so that we could depart in a barely-VFR flight. Down the highway to San Antonio we had a brisk tailwind and a rare groundspeed of 145 knots. After fueling and a late lunch, we debated entering Mexico at Monterrey with arrival after 5 p.m. or going to Brownsville for the night. The San Antonio FAA FSS had very little information on inbound Customs procedures at Monterrey, so we elected to remain in Texas.

Next morning we were off early in good weather to clear into Mexico at Tampico. Our regular Mexican tourist cards were not good for an "in-transit flight," so we each shelled out $4.50 for new cards that were turned in the following morning at Tapachula.

The Customs Inspector wanted to know in Spanish what was in that big box in the back of the 207. We told him "spare parts" and showed him what I thought was the manifest. He nodded, skimmed the two sheets of paper, quietly accepted three $1.00 bills that appeared on his desk, and then stamped the "manifest." It was only two days later, while going over some of the paper work in flight, that we discovered that our "manifest" was really the factory packing sheet for the equipment installed on the airplane. We never did find out exactly what was in that big box of parts, but our officially-stamped "manifest" seemed to work at several other border stops.

Fig. 10-6. Aerial view of Cessna's main Delivery Center in Wichita. It is from here that the authors began their 207 delivery to Santiago, Chile (courtesy Cessna Aircraft).

Then off to Vera Cruz and dreams of a super-fine seafood lunch. We were within 20 minutes of Vera Cruz, just about to start a letdown, when the tower advised another U.S. aircraft, in fairly good English, that there was no 100 octane available. However, he advised that there was 100 octane at Minatitlan. With a flurry, the charts opened up and we scrounged for our new destination. Target for the flight was another 45 minutes away, with a line of fully-developed thunderstorms along the route. Man, what a place for the weather "Flight Watch."

We stayed with the winding highway, picked up a good fix on the ADF, and bounced along in continuing moderate turbulence and rain squalls for the oil-refinery town of Minatitlan. The tower personnel spoke English and confirmed that there was 100 octane and that the gas man was there—even though this was a Sunday.

The Airport Commandante had departed and there was no one to fill out or sign a flight plan. There was also no weather information. However, there was plenty of daylight and a good highway not too far off course, so we filled out our own flight plan, had it stamped by the local Mexican airline ticket agent, and took off in a 30-knot crosswind. Legal, not quite; truly safe, not quite. The alternative was to spend the next day or two, or three, in downtown Minatitlan, and we did feel bound by the telegraphed dates of arrival in the various countries to attempt to keep on schedule.

While not hazardous, the trip across the narrow peninsula of Mexico was physically uncomfortable. As we approached the Pacific Coast Bay of Tehuantepec, surface winds whipped over the coastal ridges and the air became increasingly turbulent. The shoulder harnesses came loose and were used in front. Cameras and other unsecured bits of baggage floated around the cockpit and it was a roll-up-the-shirt-sleeves-and-drive type of flight that no one particularly enjoys. It was rough enough to remove Paul's wallet from his buttoned back pocket. Perhaps it was on this early leg that Zero One Hotel earned the name of "Big Turkey." That description continued throughout the trip, but changed from derision to sheer respect as the new Cessna kept right on plodding along over terrain that would make your palms wringing wet.

Two hours and 20 minutes later, we woke up the Tapachula tower operator from his Sunday siesta—not by radio, but only after circling the field once for a clearance to land (Fig. 10-7). The people here, as elsewhere on the trip, were courteous and friendly. We had nothing stolen and were overcharged only on rare occasions.

The Tapachula gas man said that he could sell us only one tank of fuel since everything was needed for local agricultural aircraft spraying for bugs (Fig. 10-8). The initial quote for fuel was $22. On refiguring, it computed to $13. We paid $15, and this was the best bargain in fuel on the entire trip.

The short flight to Guatamala City the next morning was relatively uneventful. For a country that spends millions catering to the tourist dollar, Guatamala's main airport is not yet geared for the general aviation pilot. We paid $3 for a landing fee, $52 in U.S. cash for 42 gallons of 100 octane. We also had to dig up our own oil spout to add the first oil to the new engine and then spent almost three hours to get back off the airport. It is truly an awe-inspiring sight to see eight (we counted 'em) officials and their helpers

Fig. 10-7. We woke up the Mexican Control Tower operator at Tapachula by circling the field late on a Sunday afternoon.

sitting watching one pair of lonesome pilots trying to get a flight plan stamped to take off. We had two sets of "local helpers" try to funnel the paper work and both finally gave up with neither success nor payment. Individually, the people were all friendly and trying to help in their own limited area; but the procedure must change with every shift of bureaucracy. If someone would paint a yellow line on the floor of the beautiful new terminal with numbered stops and a mimeographed handout sheet explaining what is needed at each stop, Guatamala City would be a tourist's delight for general aviation. As it is, you really don't care if you ever fly yourself back there, although a conducted jet-flight tour would be great.

To the best of our measuring, it was 470 nautical miles along the shoreline to San Jose, Costa Rica's capitol. The 207's "Ops Manual" lists a nautical mile range of 590 at 65 percent power at sea level, based on 73 gallons usable and a 45-minute reserve. The reserve fuel ls based on 45 percent power or 7.4 gallons, but this sort of flight planning seemed well within the ballpark. It didn't work out that way. With winds forecast at 360 degrees 10K, the planning was within reason. Also within reason was sunset, despite our forced delay in departing. The winds picked up, naturally on the nose, and as we overflew Nicaragua to avoid reported unpleasant paper work, it became increasingly evident that San Jose, darkness, and fuel exhaustion would all arrive at about the same time. Headwinds cut our groundspeed to 100-105 knots and we began shuffling our charts and books for more detailed information on Llano Grande International, our alternate.

Fig. 10-8. A shortage of fuel was experienced in Tapachula where almost all available av/gas was being used for local agricultural aircraft. It was possible to fill only one tank here.

Pre-flight briefing in the plushy offices in Guatamala City, all in Spanish, had included a marked lack of information on Llano Grande, although it was accepted as a legitimate alternate and so noted on our flight plan. Here, as in subsequent bordercrossings, the local briefers really didn't seem to care much what happened to you just as long as it didn't happen in their country.

The shadows raced quickly behind us as "Big Turkey" plodded along down the coast. We crossed a time zone, although we did not realize it until after landing. Soon, any thought of spending the night in San Jose went completely out the window and we zeroed in on Llano Grande. Our latest Sectional chart, TPC K-25C, dated 24 Nov 1972, didn't even show the airport and the adjoining road was listed as "suggested alignment." The current DoD low-altitude IFR chart supplement listed a VOR "til" frequency of 113.7. The DoD FLIP showed the airport as lighted with a "FIS" frequency of 126.7. There was also a little (*) that cross-referenced as "Operates less than continuous or on request." We gave them several calls. Nobody answered.

Two small farm strips just south of the border looked inviting as sunset crept upon us with fast, stealthy feet. We latched on to the Pan American Highway and followed the sparse line of headlights that blossomed in the greyness below. We let down to above 2500 feet, remaining high enough to spot the airport's rotating beacon—or so we hoped!

Careful thumb-on-the 1:2,000,000 scale JNC-61 dated March, 1977, eventually took us to what looked like the town of Liberia. The highway curves and crossroad checked. The road to the airport swung off to the right and we made a tight turn to stay with it. Up ahead, several clusters of lights could possibly indicate the airport, but we were so short of daylight that any probing would have been limited. It was either find that airport or land on the highway before it became too dark to see the road. Not enough fuel to make San Jose and no other alternate.

We follow down the side road religiously with a quiet cockpit and several sets of sweaty palms and straining eyeballs. Still no beacon, but like a Phoenix rising from the ashes, here come the white-painted numbers just to the north of the highway. The JNC chart, by scale, showed the field parallel to, but 3 or 4 miles from the highway. We circle just once to make sure it is the airport, drop flaps and land. Man, does that big airport look good.

We parked next to the Mooney and an older 182 and shut everything off. The first man to walk out of the office was in uniform and carrying a rifle. He didn't smile and asked in Spanish where we had come from and why we hadn't called on the radio. We assured him in "Mickey Mouse" Spanish that we had come non-stop from Guatamala City and had tried to call on the radio. Seems that things are a little touchy again tonight because there has been another border skirmish and three or four soldiers killed. Just what we needed was to be in the middle of a banana revolution.

After an indeterminate wait, one friendly immigration man shows up, checks our paper work, looks carefully at the Llano Grande printed on our Guatamalan flight plan, and finally allows that we are not unfriendlies. However, he isn't permitted to clear our baggage. As it appears the Customs man isn't coming, we take a shaving kit, it's feminine counterpart each, and he kindly drives us to a motel in Liberia.

The motel proprietor drove us to the airport in his new Toyota jeep after a leisurely breakfast and we faced the battles ahead. Fuel? There is none. Local pilot, Cecil Murray, advised that he would have gas coming in "maybe on Thursday." It was only Tuesday. Murray told us that his competitor, Manuel Guerra, about ten miles away in Filadelfia did have fuel. We confirmed this by phone, unloaded baggage and passengers before going into this short duster strip, and took off.

Sure, there was plenty of 100 octane, but it was in an underground tank—and the power was off! So the mechanics tried a handpump and it promptly broke. Next, we drained about three five-gallon cans from two parked dusters. Luckily, at this point the handpump was repaired. Since San Jose is within a hundred miles, we topped off one tank gratefully, payed the bill with a hefty tip to a group of very nice Costa Ricans, and headed back to Llano Grande International (Fig. 10-9, Fig. 10-10 and Fig. 10-11).

Anticipation of further problems in paper work, etc., proves groundless for once. With everyone aboard, we headed up the highway and work San Jose approach into an uneventful landing at the broad Santa Maria Airport.

The next morning, refreshed and certain that all the problems are behind us, we depart for David and Panama City. We could have made Panama City's Paitilla Airport without the stop at David, but remembering Guatamala City we felt that the paper work would be easier and bordercrossing less complex at a smaller airport. At David the paper work was less, but the only person on the airport with a key to the gas pit at the agricultural FBO was out

Fig. 10-9. Mechanics at Filadelfia, Costa Rica, attempt to get 100-octane fuel from a buried tank with a hand wobble pump. The pump broke. Cessna 207 is in the background.

Fig. 10-10. Here's one way to get 100 octane in Costa Rica. Filter it out of the wing tanks of a Snow agricultural airplane into plastic five-gallon containers.

Fig. 10-11. 100 octane finally flows into the tank (singular) of the Cessna at Filadelfia.

for lunch and siesta and would be back "at about 2 p.m." It was then noon. We would have crowded our existing fuel to continue without topping off. Someone commented that "love is a full tank of gas." We waited and waited and waited. We tried to bet the mechanic 10 to 1 that the key holder would be late. He would have lost.

In Panama City, Paitilla is a comfortable, efficient downtown airport—only $1 by cab to the center of the city. This busy airport is for the commuters and general aviation aircraft on the Pacific side of the Canal Zone. Inbound paper work was at a minimum since our passports had been stamped at David. We off-loaded baggage and headed for the Caribe Hotel, recommended in the AAA's travel guide.

It was time for an oil change. Our searching led to Hector Reyes, chief mechanic for Aerolinas Perlas (three twin-Otters and one Islander). Senor Reyes doesn't speak much English, but if you're ever in Panama and need any mechanical work, look him up (Fig. 10-12). The closest Cessna dealer is in Costa Rica.

In Panama, we changed $350 in Bank of America travelers checks to U.S. dollars. Throughout the trip we found banks to be open only in the morning. Neither the local hotel lobby nor anyone else would handle the transaction—only the bank. Despite the best efforts of Karl Malden and his TV commercials, we found travelers checks to be much less than satisfactory. They're fine to pay hotel or food bills, but usually unacceptable for gas, landing fees and the other steady stream of unforeseen expenses. In all of Central America and as far south as Peru, our Diners Club Card was excellent, but the only true common denominator was U.S. folding money.

All along the way, food and local foamy beverages were excellent. None of us had Montezuma's revenge on the entire trip, although we used certain restraint with unpeeled vegetables and some tap water.

The FAA in the Canal Zone indicated that we could expect an excellent weather briefing merely by a phone call to Base Operations at Howard Air Base prior to going to the airport. The briefing was fine, but completely inaccurate. Since our non-turbocharged 207 showed a service ceiling of only 13,000 feet, we elected to bypass our planned fuel stop at Medellin, Colombia, elevation 8225 feet and ringed by "extremely high mountains in immediate vicinity" (stated on the Colombian chart). The low altitude supplement gives 17,000 feet as the lowest en route IFR altitude.

The logical low-altitude airport of entry in Colombia is Cali, where it is lonesome coastline and jungle all the way down.

Any flight over, or landing in, a Central or South American country requires advance notice of date and place. Some countries also require permission. We had availed ourselves of AOPA's Flight Department service for notification by telegram and payment of the $30 overflight fee for each of the four South American countries on our route. Expecting a turbocharged 207 from the factory, our telegraphed flight plan scheduled us to fuel at Medellin and then overnight in Quito, Ecuador. From there to Tulara, Peru for fuel and on to Lima where we were planning to "vacation" for three days, including a side trip to Cuzco. Arequipa was to be our last stop in Peru before crossing the border at Arica, Chile and continuing southeast.

Fig. 10-12. Chief Mechanic Hector Reyes of Aerolinas Perlas changes the oil at 25 hours on the author's Cessna 207. Specified break-in mineral oil was extremely difficult to find on the Paitilla Airport.

By now we were three days behind our telegraphed arrival date in Colombia and we have again changed our destination airport—as we had twice before in Central America. But so far, no one has raised any objections (Fig. 10-13 and Fig. 10-14).

The departure fee and overnight parking at Paitilla Airport came to $18.50. The combination tower operators/flight service technicians accepted our three-hour, twenty-minute flight plan to Cali without question. We listed Buenaventura, Colombia (marked as international) as an alternate, and we were on our way.

After takeoff, we cleared through the control zone at Tocumen International, bypassed the Pan American Highway listed as "under construction" and headed down the circuitous coastline toward the VOR at La Palma, Jaque and points south. After our tailbearing from Panama's Taboga Island VOR sank like the sunset, we tuned La Palma and nothing happened. We relayed through a Pan American Clipper and were eventually advised that the VOR was off the air. Such a detail might have been handy to know before takeoff.

Our VFR chart, TPC-K26D, dated six years ago, was the latest available large-scale chart of this area. The shoreline along this "Ensenada de Garachine" and the coastal side of the low Serrina del Xanazas hills is some of the most uninviting territory in the world. It is jungle and swamp without roads. The occasional thatched-roof village—three huts and two dugout canoes—were even few and far between. We tried not to think about the assorted reptiles, bugs, animals and jungle rot that probably lay in waiting under that green mat of tree tops sprinkled with the occasional reflection of swampy water. Disneyland's jungle ride this is not.

The Cordillera Occidental coastal range, shown on the old charts, never did come into sight in the haze and scattered fog. As we approached Cali, the cumulus cloud build-ups increased. It would have been nice to have received a fresh terminal weather report for Cali, but our calls produced nothing but rapid-fire Spanish followed with a succinct "goodby." We contacted several Spanish-speaking air carriers working Cali center, but usable information was zilch.

However, it really didn't make much difference because the cloud tops were reaching perhaps 20,000 feet above the coastal haze. We looked longingly toward the pass—complete with railroad and main highway—that extended from Colombia's one seaport, Buenaventura, in toward Cali. With bases at about 1500 feet and visibility dropping below three miles, there was no way to explore

Fig. 10-13. Cessna 207 awaits passengers and crew beside a short Skyvan at Panama City's Paitilla Airport.

that alternate. We'd already been in the air over three hours and didn't have that much fuel for serious exploring.

We had a firm fix on Buenaventura's NDB and headed for our alternate. This was another in a succession of unplanned stops, but there really wasn't any option. Remember that fine forecast from Howard Air Base in the Canal Zone!

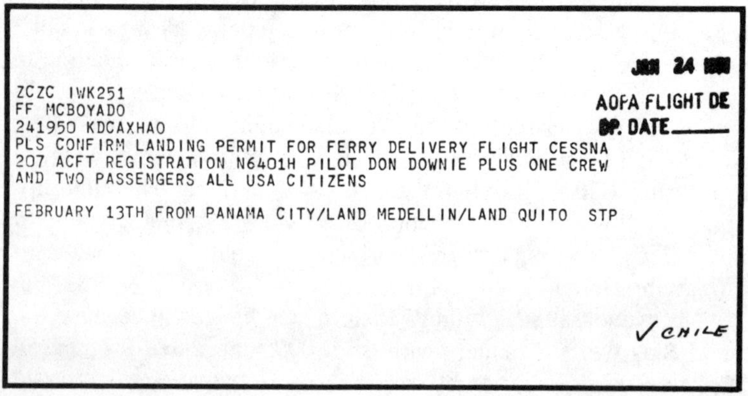

JAN 24

AOPA FLIGHT DE

DP. DATE

ZCZC IWK251
FF MCBOYADO
241950 KDCAXHAO
PLS CONFIRM LANDING PERMIT FOR FERRY DELIVERY FLIGHT CESSNA 207 ACFT REGISTRATION N6401H PILOT DON DOWNIE PLUS ONE CREW AND TWO PASSENGERS ALL USA CITIZENS

FEBRUARY 13TH FROM PANAMA CITY/LAND MEDELLIN/LAND QUITO STP

✓ CHILE

Fig. 10-14. AOPA's telegraphic communication requests permission to fly from Panama City to Medellin, Colombia, on February 13th. The flight actually took place on March 2nd.

Wouldn't it have been a simple, logical solution to land, top off fuel tanks and continue on down the coastline? Guess again, you're now in South America and there are more borders to cross.

We chased down the NDB needle and soon there was a break in the jungle beside yet another winding river. The oiled flight strip at Buenaventura International looked singularly inviting under these conditions (Fig. 10-15). We raised the tower operator who tried his best to give instructions in what would never pass as English. We circled and landed.

The tower operator smiled, closed our flight plan and advised that there was no fuel service at this airport. He was apologetic when he detailed that the Colombian government, in an effort to crack down on drug smugglers, had curtailed the sale of all av/gas. None of the fuel depots in Buenaventura had av/gas. The nearest fuel was at Cali, and a pilot would have to sign for it. That was at least three hours each way by pickup truck plus the time of tracking down a truck and the fuel (Fig. 10-16).

Fodor's Guide to South America lists Buenaventura as "the country's busiest harbor, which is mainly a port of entry. There is not much here to induce a traveler to linger here." End quote. The travel statement might be well taken, but again the people were cooperative.

The logistics of a three-hour trip plus the inevitable hassle at the airport indicated a full day just to get fuel. We picked up the yellow pages and had Senor Gallego, the hotel proprietor, call the Cali Piper Dealer, Aero Occidente, where we finally found someone with a fair knowledge of English. Yes, they could ferry fuel to us. The price would be 5000 pesos ($200 per hour) for the flight in their 206 plus the cost of the fuel. How much fuel did we need? Fifty to 60 gallons. Yes, Captain Jaime Santamaria would fly the fuel down just as soon as their 206, No. 1444, was out of the shop. That might be a couple of hours. Fine, we'd meet at the airport about 10:30.

What do you know, it worked! By not later than 11:00 a.m., the 206 called Buenaventura Tower and soon rolled up beside our 207. Captain Santamaria confirmed that there would have been no way for us to get into Cali VFR the previous day or today (Fig. 10-17). He had two helpers with five-gallon plastic jerry cans, and within 10 minutes our tanks were full. Captain Santamaria looked at the new 207 and commented that it wouldn't do him much good without a turbo. He took 8000 pesos and gave me a receipt before heading back to Cali (Fig. 10-18 and Fig. 10-19).

Fig. 10-15. Buenaventura "International" Airport, as seen from the Cessna 207 prior to landing under marginal weather conditions. NDB was useful in finding the airport.

The Airport Commandante/radio operator collected $40 for use of the International Airport—he showed us in his books where this fee was required of foreign aircraft on their initial landing in Colombia. He checked on the radio to assure that fuel was available at Tumaco, another 175 nm down the muddy coastline or over solid jungle. We knew that we could reach Guayaquil, Ecuador from Tumaco, Colombia, bypassing Esmeraldas, Ecuador and an unknown fuel availability.

Fig. 10-16. Luggage comes out of the 207 at Buenaventura International prior to a very bumpy ride into town with a reluctant cab driver. Left to right are Julia Downie, Paul and June Crawford.

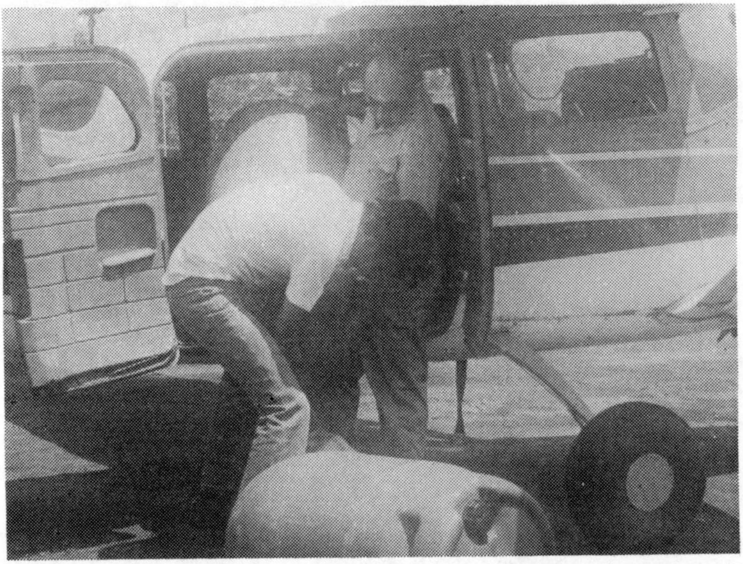

Fig. 10-17. Capitan Santamaria and a helper empty a 30-gallon plastic container of fuel into five-gallon containers at Buenaventura, Colombia.

Fig. 10-18. Paul and June Crawford watch refueling from a Turbo 206 into the 207 at Buenaventura, Colombia. Tower operator in the background monitors the operation.

Fig. 10-19. Not all the expensive 100-octane fuel reaches our tanks as young, enthusiastic Colombians pour from a five-gallon can through a funnel into the long-range tanks of the 207.

At Tumaco, no one answered the radio, although we could see many cars parked around the terminal. We landed and rolled up in front of the terminal to be met by at least a dozen young men in uniform (Fig. 10-20). Most carried automatic rifles. This junket would certainly have an added interest to members of the National Rifle Association. The airport commandante appeared, looked

Fig. 10-20. Young soldiers with truncheons surround the 207 after landing at Tumaco, Colombia. These troops were courteous and inquisitive.

with some suspicion at our flight plan from Buenaventura and finally allowed that we could fuel and go on. There was a single, rusty 55-gallon drum, but the fuel color was greenish (Fig. 10-21). We topped only one tank—28 gallons, including what was spilled. The cost—$3.00 per gallon! We finally got an unsigned typewritten receipt, but departed knowing that we'd been "had." For the remainder of the trip, the town was called "Too-Much-O."

Inbound to the Ecuadorial capitol, we passed over lush green farms on low hillsides that were probably coffee. Thatched roofed huts and towering palm trees made a really spectacular setting despite the hazy overcast and light rain that made picture-taking less than satisfactory. Then came an area down the Rio Grande de Daule for the last 30 to 40 miles of rice paddies with houses on stilts isolated in the middle of man-made lagoons. ONC-M25 dated six years earlier had no relief information. On a clear day, it would have been outstanding.

Finding Simon Bolivar Airport was a mixture of pilotage and IFR procedures. The strong Palma NBM gave us an indication of the airport and we worked approach for the final 20 miles, reporting over the NDB with the airport still not in sight. So you fly the inbound heading of 206 degrees for the instrument approach and finally the 8005' runway appears out of the haze.

This was the first stop where a check was made of our telegraphic permission to land in the country. Even though our stateside flight planning had listed Quito as a port of entry several days earlier, the flight office in Guayaquil seemed not bothered by either the change in plans nor the inaccuracy of the entry date. We appreciated this laxity, but it helped set us up for the next leg of our flight.

After encountering continuing problems with fuel and customs at the smaller towns, we picked the largest town by size of area on the map within range in Peru and filed for Chiclayo. That was fine with the Ecuadorian flight planning desk, so we took off—fat, dumb and happy. Weather had been forecast severe clear and we were almost IFR with low clouds four minutes after takeoff. The first VOR on our charts at Talara wasn't on the air. But that was only the beginning. We crossed the border and then headed across Peru's large desert area that looked a little like the sands south of the Mediterranian, but not as vast.

Chiclayo's VOR came in strongly and soon we had the field in sight. We called the tower and the reply was in rapid Spanish. The only word that stood out was "prohibido." Suddenly the airport looked very military, although there was an air carrier terminal on the west side of the runway. In inadequate Spanish we relayed that we didn't have the fuel to continue on. Finally the tower said "cleared to land," and we were in trouble again.

Not only had we landed at a military base without authorization—even though none of the charts had it marked as such and the Ecuadorian flight planning people had blissfully okayed the flight plan—but we'd also entered Peru on the wrong day according to our telegraphed flight plan.

It was a long, hot afternoon with phone calls and wires to Lima. Finally, at about the time we just had enough daylight to make Lima, a wire came through authorizing our takeoff, but we would have to report to the Ministero of Transportes, Senor Technico Vela, the following day.

Lima is a beautiful city with crazy cab drivers, fine parks and great inexpensive food. However, we really didn't appreciate any of that as we walked up the steps of the Governmental Transport building. Senor Vega passed us along to a very nice "doctore" who spoke no English and there was no interpreter available. There was, however, a fairly fat stack of papers pertaining to Cessna N6401H. We spent the next two and a half hours while he prepared a page and a half deposition in Spanish explaining why we had broken Peruvian laws, that we were sorry and had made these transgressions out of ignorance.

The deposition was typed, we reluctantly signed it—there was no option—and the "doctore" disappeared into the Minister's office. He returned after an indeterminate time and smiled. Everything was taken care of, or would be just as soon as we paid a $75 fine (Fig. 10-22). It was just before lunch and siesta, but we were able to get a receipt and a signed piece of paper that said we were now legal to continue our flight through Peru. Then we walked out into the beautiful day.

After our time-and-money-consuming experience with "the Feds" in Lima, we considered it prudent to send another arrival telegram to Chile. You'd think that the big, Jose Chavez International Airport would have telegraph or TWX facilities. No way—you drive downtown, this time in company with one of the flight planning staff who would provide transportation and know-how in exchange for a tank of gas in his VW. The telegram to Santiago advising that we would be crossing the border within three days cost $17. Originally we'd written the notification that we'd arrive

Fig. 10-21. Fuel from this rusty drum at Tumaco, Colombia, cost $3.00 (U.S.) per gallon and an unsigned receipt was all that was available. A rip-off? Perhaps, but what can you do when you want to get the heck out of the Country?

Resolución Directoral N°0013-79 TC/AE.ca

Lima, 16 de Marzo de 1979

Visto el Dictamen de la Junta de Infracciones de 7 de Marzo de 1979 relacionado con el ingreso sin autorización al territorio nacional de la aeronave Cessna 207 de matrícula norteamericana N-6401-H.

CONSIDERANDO:

Que el día 5 de Marzo de 1979 ingresó al territorio nacional, aterrizando en el Aeropuerto de Chiclayo, la aeronave Cessna 207 de matrícula N-6401-H, al mando del piloto de nacionalidad norteamericana, señor Donald Chase Downie, sin contar con el respectivo Permiso de Vuelo, en razón de encontrarse vencido el anteriormente otorgado, infringiendo por lo tanto la reglamentación sobre tránsito de aeronaves de matrícula extranjera en el territorio nacional.

Estando a lo previsto en el artículo 2.1 inciso 5) del Reglamento de Sanciones, Anexo al Reglamento de la Ley 15720.

SE RESUELVE:

ARTICULO UNICO.- SANCIONAR con multa de QUINCE MIL SOLES ORO (S/.15,000.00), al piloto de nacionalidad norteamericana, señor Donald Chase Downie por su responsabilidad a que se refiere la parte considerativa de la presente Resolución Directoral.

Regístrese y Comuníquese

El Coronel FAP.
LUIS IRIARTE VENEGAS
Director General de Transporte Aéreo

Fig. 10-22. This is what a $75 (U.S.) fine looks like in Peru. If you can't speak and read Spanish, you have a problem.

the next day, but our local flight planning expert, who proved smarter than we were, suggested we give a couple of days leeway.

The next morning our passengers decided to stay in Peru to visit friends from forty years ago and we parted company on a busy one-way street in Lima. At the airport, there was no gas truck; you must taxi from the transient parking area to fuel and return to parking. Then comes the flight planning, weather report and the everpresent fees—$27.95 for airport use. There's even a $10 per head departure tax for airline passengers.

Southeast of Lima, there wasn't an airport within safe range at which we were permitted to land other than Arequipa, 325 nm and elevation approximately 8000 feet. Tacna, the official international airport closest to Chile was 450 nm, and with the headwinds we had been consistently experiencing, we had doubts about fuel (Fig. 10-23).

Arequipa was foggy when we filed our flight plan, but the forecast for clearing by noon was bolstered by the profound pronouncements from the Lima meterological staff that "it always clears up there by noon." It didn't!

At Lima, we were advised to avoid the charted restricted area and fly well offshore at Pisco. Aside from this detour, the leg down the coastline of Peru was comfortable, relatively civilized with occasional small towns and tiny agricultural flight strips. Weather off our left wing continued to obscure the Andes that showed plainly on our cockpit charts. With a minor headwind we plodded along at 120 nm.

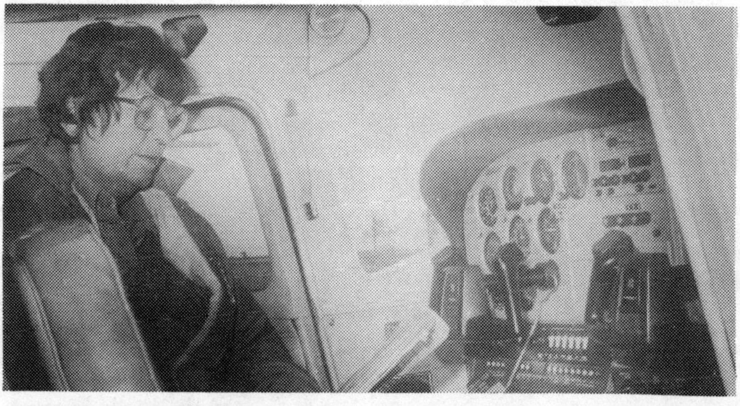

Fig. 10-23. Julia Downie checks the international paper work in the cockpit of the Cessna 207 prior to takeoff.

We turned inland 50 miles from Mollendo to avoid another restricted area and followed a broad, paved highway toward Arequipa. The NDB came on comfortably, but its identification code was hard to read at a distance.

ONC P-26 lists a little blue circle for an airport called Vitor, 6200 feet long at 5167 feet altitude. When a long, combination oiled and gravel strip came below the windows of the 207, we dutifully checked off Vitor. Within about three minutes came a huge paved flight strip, hangars, assorted buildings—but no cars, no airplanes and no signs of people. Our ADF needle bird-dogged in at 45 degrees off the left wing and the town, located in the proper direction from the airport, didn't look large enough for Arequipa. Why don't the chart makers put runway headings on the VFR charts for the few paved runways in this part of the world.

In any event, after our misadventures at Chiclayo, we surveyed the big airport carefully, called tower/approach and didn't get an immediate answer. So, we headed back to the main highway, followed the ADF bearing and finally wound our way in a drizzly rain and a ceiling that descended as we climbed, until we were able to talk with the Arequipa Airport (Fig. 10-24). Again there was no English, but we were able to discern that visibility was six kilometers with light rain.

We continued on up a fairly broad winding canyon, following the highway, and on to the outskirts of what was really Arequipa. There was a great deal of eyeball-searching through the haze as we crossed town and followed both road and NDB to the airport. Our first glimpse was a bright rotating beacon in the drizzle. So the airport's below VFR minimums—but to heck with that. They cleared us to land and we did.

We were allowed to fuel, but not to take off, as the field was below minimums. After another unplanned overnight, we took a cab to the airport the next morning without much chance of getting into the air from the looks of the weather. "Yes, the airport, she was still closed." When might it open? Maybe later today, maybe tomorrow. A week ago, it hadn't opened for seven days. And at this point, the airport commandante advised that he couldn't permit us to take off anyway as we hadn't landed on the day shown in the telegram in his files—we were three days late. Our flight plan from Lima to Arequipa had no importance. It took the day before's date on our receipt for paying the fine in Lima plus a phone call to Lima to assure him we were legal. Right at this spot, the glamour of far away places via small airplanes had finally lost most of its luster.

Fig. 10-24. In-flight photo of the Cessna 207 at 9,000 feet approaching Arequipa, Peru. Note minimum nav/com equipment. Both ADF and VOR operated perfectly on this delivery.

With nothing else to do, we waited around the airport, shot the breeze with a government mapping crew whose Learjet was also awaiting better weather, and watched the people in the terminal (Fig. 10-26 and Fig. 10-27). Baggage began collecting on the carts and word got around that the 727 might make it today. We went to the Commandante again, and he smiled and said that we'd be able to take off in about 40 minutes. The weather really wasn't any better, but we quickly made out a flight plan for Arica, Chile, only 175 miles distant, and paid the 65 cents landing fee.

Gradually, it became apparent that the airport would be VFR as soon as the air carrier jet landed, regardless of the weather.

Fig. 10-25. Faucett airplane, designed for flying the Andes, is parked in the middle of the airport at Arequipa. Built in Brazil, this historical airplane should go into a museum.

Fig. 10-26. A young Peruvian looks at the big single-engine Faucett at the airport in Arequipa. The new airport terminal can be seen in the background.

Rather than push the situation, we watched hoards of people come off and then another crowd board the jet, and we fired up the 207 only as the three-holer taxied out. The minute he cleared the runway, we were ready to go and cleared.

It was downhill all the way and one hour and 20 minutes later we were talking excellent English with the Chilean tower operator at Arica.

By the time we had waded through the paper work (Fig. 10-28), topped our tanks, arranged for payment and prepared to go, we were within a couple of hours of darkness and faced a 400 plus nm flight to Antofagasta. We decided not to push it and took a long cab ride—17 miles—to town.

Things Are Looking Up

The Chilean officials the next morning were cooperative and re-issued the flight plan to Antofagasta and associated paper work without extra charge. Things really began looking up. We were told to file the W200 airway that took us 20 miles off shore and around a large restricted area at Iquique. We complied and then swung back within landing distance of the baren shoreline as we zero'd in on Antofagasta where, again, both approach and tower personnel spoke excellent English. We found out later that this was a national requirement in Chile for both tower and en route controllers to speak fluent English or they were out of a job.

The broad, General Cerro Moreno Airport in Antofagasta was a joint-use airport with a large military contingent on the north end of the airport. We landed straight in and taxied up to the big, modern terminal and . tower. Again, we were met by an Air Force Lieutenant with a gun who was unhappy about our flying over a prohibited area. Here we go again! Our VFR charts were current—only 8 months old—and showed no restricted area. We went to the combination tower/approach and the English-speaking operator said that the military was complaining that we had violated a newly created prohibited area. The tower operator was obviously on our side and called back to Arica to verify that our clearance had been to fly offshore over Pisaqua and Iquique and then continue to Antofagasta. Later—after the Air Force man left—the tower operator explained that the new prohibited area had been put in recently by the military without any provision to get from the offshore airway to the airport. All low-level flights by foreign aircraft were required to use the off-shore airway.

When we went to fuel, another unexpected problem arose. They would not accept Chilean money in payment as the aircraft was U.S. registry and only U.S. dollars were acceptable. It seems that general aviation aircraft in Chile buy fuel coupons from their Air Force and then use the coupons at the gas pumps. The gas man reluctantly said that he would take travelers checks for fuel but he would not be able to provide any change. Wouldn't you guess that the total was $41.75. For once the odds worked in our favor as we signed two $20 travelers checks and paid the remainder with U.S. silver quarters—the last of our change.

We had been told to take up a heading of 200 degrees after takeoff and stay with that until we were 20 miles off shore. Then change to a heading of 180 degrees until passing Caldera 200 plus miles down the coastline. It was wet and lonesome, but those were the orders. The big Continental went into automatic rough, psychologically, but continued to do its job admirably.

Just after takeoff, the tower called and asked for the number of our landing receipt from Arica. We were almost tempted to turn the radio off and just keep going,. but finally found the receipt and relayed the number. Only later did we find out that this $26 receipt covers landing fees for any airport in Chile for visiting aircraft for the next 30 days.

Fig. 10-27. Fabric damage on the Faucett at the Arequipa Airport indicates that it will be a useless memory unless steps are taken to restore the airplane.

Weather ahead went from scattered to broken to virtually solid as we droned along at 8000 feet. With no way of finding out what the weather was really like ahead and no instrument approach chart for our destination of La Serena, we dropped down through one of the last remaining breaks in the clouds and leveled off at 800 feet above the water. Shore was a long way away and we hedged on our 20-mile restriction to keep the shoreline in sight. Not knowing the accuracy of Chilean radar, we sort of compromised at low altitude and 4 or 5 miles off shore.

Once past Caldera, we picked up the beautiful shoreline with its paved highway and continued on to La Serena where the tower operator again spoke good English. We waited in dread anticipation with partially baited breath for the "ax to fall" regarding our departure from the 20-mile-out instruction, but nothing was ever

JAN 26 1978

AOPA FLIGHT DEPT.
OP. DATE________

A.K NNNNZCZC WIK357 262044
JJ KDCAXHAO SCARYD SCARZP SCARZH SCIQYD SCIQZP SCIQZH SCSEYD SCSEZP
261937 SCLEYAVG
NR 10/1/127 CLR ARR/SOBREV ACFT CESSNA 207 COMA MATRICULA
N5401H PTO PLT DON DOWNIE PTO 1 TRI PTO 2 PAXS PTO FECHA Y RUTA
18 FEBRERO AREQUIPA/A1/SCAR/W200/IQQCAVANCHA/W200/SCFA/W200/CLR/
A1/SCSE PTO 19 FEBRERO SCSE/UA1/TOY/UA5/SCEL/ NEGATIVO ARR EN
CALDERA PTO VUELO FERRY PTO FVR INFO ARR/DEP PTO

Fig. 10-28. Reply to AOPA telegram for permission to fly into Chile. The permit calls for landing on February 18th, but prohibits landing at Caldera. A working knowledge of Spanish is essential here.

said. The weather was reported good and we had daylight enough to make Santiago, so let's fuel and go. No way! The gas man wasn't on the field and might be back in an hour or so. We'd been that route before, so we checked on a hotel.

La Serena lived up to its name. We didn't see a gun or a uniform on the entire field. The tower operator gave us a phone number to check weather the next morning and prophecied correctly that there would be some early morning stratus.

In the morning, the gas man was there and would take our travelers checks; again—no change, but we came out fairly close. We filed our flight plan to Santiago. Weather was hazy but good and the paved highway pointed in the right direction. As we taxied out, a sport parachutist drifted down for an easy landing. At long last, we could see the skyline of the Andes as we cut the corner abeam Valpariso and picked up a good ADF fix on Santiago.

As we attempted to raise Santiago Approach, a Braniff jet en route to Buenos Aires, Brazil, was having the same problem. We talked briefly and the air carrier captain wanted to know how long it had taken us from Wichita. Forty-six hours and 17 days. His English sounded good and "stateside" on the air.

Aeroservicios' owner, Jose Vidal, was in Antofagasta with a 310, so his chief mechanic Roberto Saldiria signed our much-worn aircraft manifest as a receipt for the delivery, and we called Braniff for standby space on that night's flight back to Los Angeles. It was only after riding for more than 12 hours in the Stretched DC-8 to get back to our starting point that we realized just how long in miles the trip had been.

In retrospect, neither of us would volunteer to take another non-turbo, single-engine delivery all the way to South America. For someone else contemplating the trip, we'd seriously recommend "tanking" the ship for long hops off-shore with the associated life rafts and ditching gear. Properly equipped, a landing in open water would seem preferable to some of the jungles south of Panama.

As political events change and light airplane travel becomes a more accepted way of life south of the Panama Canal, this flight might be a satisfying experience. However, until that time, we'd recommend anything south of Panama to the true professionals who have access to current border requirements and will accept the risks that go with flying in primitive areas for the first time.

Should a flight come up to Panama tomorrow, well—why not? Beyond that—why?

Index